HOW BUILDINGS WORK

HUW M. A. EVANS

RIBA Publishing

How Buildings Work

Published by RIBA Publishing, part of RIBA Enterprises Ltd, The Old Post Office, St Nicholas Street, Newcastle upon Tyne, NE1 1RH

ISBN 978 1 85946 557 8 / 978 1 85946 752 7 (Ebook)

British Library Cataloguing-in-Publication Data
A catalogue record for this book is available from the British Library.

Commissioning Editor: Fay Gibbons
Production: Richard Blackburn
Design and typeset: Kalina Norton, Studio Kalinka
Printed and bound by W&G Baird Ltd in Great Britain
Cover image: Anna Mill
Figures drawn by Neil Storey from Huw Evans' sketches

www.ribaenterprises.com

Contents

About the author

Huw M A Evans has worked as a technical author and trainer in the construction industry since 1996 and holds a Postgraduate Diploma in Technical Authorship from Sheffield Hallam University. From 1996 to 2006 he worked at JPA Technical Literature in Newcastle-upon-Tyne, developing technical documentation both for the UK and Europe, and providing technical support for thermal calculation software.

Huw has been a freelance consultant (www.writelines.biz) since 2006, authoring technical documents and writing and delivering training courses for designers, energy assessors and construction product manufacturers. He is also the author of the hugely popular *Guide to the Building Regulations*, now in its 3rd edition. Huw has particular interests in building physics and energy efficiency, and how they are addressed by building regulations.

Acknowledgements

I must begin by expressing my appreciation of Eric Breeze, one of my secondary school physics teachers, whose slogan was 'physics is fun': it turns out he was right. I trust he would have been gratified to learn that, decades on, one of his pupils was sufficiently engaged with physics to write a book about it.

Next, I must thank John Potter for introducing me first to the application of the theory of physics to the practical matters of designing and constructing buildings, and secondly to the need to present the recommendations as clearly as possible.

Fay Gibbons, the commissioning editor for this book, has been a great help, particularly in the structuring of the text. Her clear-eyed view of the wood has been invaluable while I have been grubbing at the roots of the trees. However, the fact that I am constitutionally incapable of adopting other people's suggested structures ensures that infelicities of organisation remain my own.

I would also like to thank Dr Claire Das Bhaumik (Inkling LLP), Matthew Frost and Huw Heywood for reviewing early drafts of the manuscript: their comments and suggestions have substantially improved the final text.

My greatest thanks are reserved for my family, who have supported me throughout the writing process, particularly during the final months when the whole of life seemed wrapped in a steadily thickening fog of building physics.

HMAE
January 2016

Preface

An understanding of building physics is, I suggest, fundamental for the design and construction of safe, functional, energy-efficient buildings. We need to understand how buildings work – not in the sense of circulation spaces, massing and the like but rather how a building will perform: for example how much ventilation will be required for the likely occupants and how can that be provided without compromising the thermal performance; or how will the layout and fenestration affect light levels and cooling demand.

My intention in this book is to give an overview of the main physical phenomena that affect the use and operation of buildings and to explore those effects and their interactions, but always with the aim of addressing the usability of buildings.

At this point I have to acknowledge that the term 'building physics' can be off-putting: it may have unwelcome associations with tedious experiments at school with mirrors and pins, inexplicable circuits or trolleys running down slopes. For the ancient Greeks *physis* – from which we get our word physics – was the whole of nature (which included not just the physical world – there's the word again – but also beings and phenomena which we would classify as 'spiritual' or 'supernatural'). Of course, academic subjects have divided and mutated over the years, and our understanding of the cosmos has changed, but physics remains a way of understanding the interactions of the materials and forces of which nature is composed.

And building physics is a division within that larger subject which seeks to understand how a building is affected by – and affects – its environment and its occupants. Perhaps the best way to explore that is to consider a building with which I am very familiar: my house.

It was built in 1925, with masonry cavity walls (with only air and wall ties in the cavities), a tiled roof with neither insulation nor underlay, and timber floors with a very well-ventilated void beneath. The outhouse for the laundry was roofed, but not entirely enclosed. The wood windows were single glazed, with opening casements. There were four fireplaces, and a kitchen range which also heated water for the bathroom.

It was not old-fashioned though: it had electric lighting from the start (in comparison, my father's home town did not get mains electricity until about 1930) and the long greenhouse at the end of the garden was centrally heated by 4 inch diameter hot water pipes.

In the 90 years since it was built the house has been adapted and remodelled: the wall cavities have been filled with blown-fibre insulation; the roof has been re-covered and insulated, and most of the windows have been replaced with double-glazed units in frames made of synthetic material (uPVC). The fireplaces have been blocked up and central heating fitted; the house has been rewired twice and the laundry has become an enclosed utility room.

Those changes occurred not by random mutation, but as the result of deliberate (and frequently expensive) choices on the part of successive householders. Choices made, in part, on the basis of changing expectations of what a house should provide. Why be too hot in a room with a blazing fire and yet too cold elsewhere, when a central heating system can heat the whole house? And once that is done, why put up with the gusty draughts drawn up the chimneys?

Of course, there have been other motivations. Householders responded to the energy disruptions and price shocks of the 1970s by improving energy efficiency; the first few inches of roof insulation probably date from that decade. Snow drifted in through the leading of the windows and the traffic noise became increasingly intrusive: the wooden windows went, replaced by 'maintenance-free' uPVC. Even the cutting-edge incandescent lamps have been replaced by compact fluorescent lamps and light-emitting diodes (LEDs).

Despite all the changes the house still fulfils its original functions of keeping the occupants safe, dry and (reasonably) warm. And, despite all the changes, the fundamental physical processes that affect and are affected by the house operate as before: electromagnetic radiation from the sun passes through the glazing of the windows, lighting and warming the rooms; and large-scale air movement creates pressure differentials across the house, which in turn produce draughts.

Yet, although the 'laws of physics' are unchanged (even if our understanding of them has sharpened since the 1920s) the scope of their operation has changed. Rooms are heated by convection currents generated by central heating radiators, instead of the radiant heat from the fires. We probably wear fewer layers of clothing around the house than the first occupants, and keep a larger proportion of it at a higher temperature. There is substantially less air movement: there is no updraught generated by the chimneys, the new windows are well-sealed and – because we are using deodorant, washing more frequently and not burning coal – we do not fling the windows wide for fresh air. With less air movement the water vapour generated by bathing, showering (the shower is new) and laundry moves through the house and passes into the fabric by air leakage and diffusion, rather than being carried rapidly out of the house.

In addition, the behaviour of the building's occupants has changed and any analysis of a building's physics has to consider those occupants, because buildings have to work for people.

In Arthur C Clarke's novel *Childhood's End* the last man stows away on an alien spaceship to reach the aliens' home planet. He has a miserable time there, in part because of what he discovers about humanity's fate, but also because the alien buildings are made to suit the aliens' abilities, one of which is flight: the stowaway's explorations are curtailed by an absence of stairs.

We are used to stairs being the right size for human use, which is something that has been learned and eventually codified. But when it comes to other things we are less accustomed to acknowledging the human measure. In the huge range of temperatures that the universe offers we can survive in only a narrow range and are comfortable in an even narrower one. Our eyes can perceive a limited band of the electromagnetic spectrum, and exposure to other parts of the spectrum is harmful. We can see by sunlight and moonlight, but not by starlight.

Not surprisingly, we construct and operate our buildings to address those limitations as best we can. I am not primarily thinking of questions of why buildings are the way they are: there are plenty of books dealing with power structures, cultural imperatives and aesthetics. I am interested in the fundamentals of shelter, security and wellbeing, which we strive to achieve through the interaction of three things: human physiology, buildings and building physics, which is illustrated in the 'building physics triangle'.

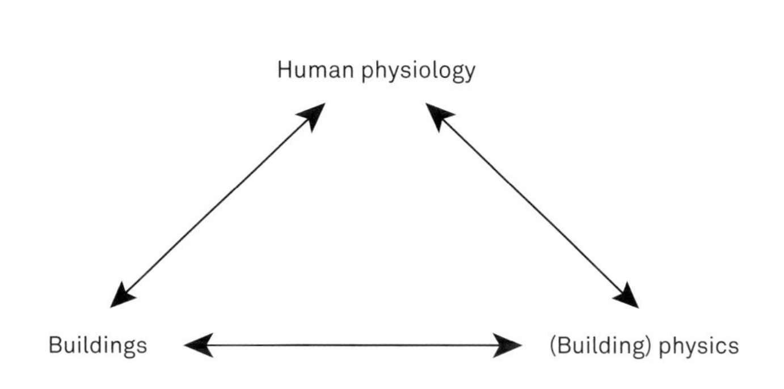

THE BUILDING PHYSICS TRIANGLE

That triangle of relationships underpins the rest of this book: it applies to a cabin in the Florida panhandle, a concert hall in Norway or a museum in Japan.

Looking at the first point of the triangle we can appreciate that, while the expectations of the cabin-dweller, concert-goer or museum visitor may differ, they are subject to the same physiological constraints. On the second point, as Scotty observed in *Star Trek*, 'we cannae change the laws of physics' which (in a descriptive–predictive sense) are the same across the planet. Air above Australia is affected by forces in the same way as the air above Alaska: in either place, if air that is already saturated with water vapour is cooled the result will be condensation. Which brings us to the third point of the triangle and the real challenge: designing buildings. That is the justification for looking at building physics.

But in thinking and writing about building physics we face the problem that occurs whenever we think and write about complex, interrelated phenomena: we have to divide them up to consider them one at a time. To write about the principles of heat transfer I have to isolate it from other phenomena such as moisture, even though I know that moisture content has an effect on heat transfer. Simply to get text into meaningful chapters I have to draw lines around subjects, deciding that *this* belongs in *Light* and *that* belongs in *Heat*, yet at the same time knowing it all happens at the same time. (Also, a book, unlike physical phenomena, has to have a boundary: for reasons of space some interesting topics, such as thermal movement, have had to be excluded.)

As a result, each chapter focuses on one of the five primary physical phenomena: heat, air, moisture, sound and light. (Key terms, given in bold, are defined in the Glossary pp. 134–37.) At the end of each chapter is a diagram that shows the main points of interaction between the subject of the chapter and the other phenomena, and identifies the main design implications of those interactions. If the diagrams do nothing else they will remind us that to think successfully about building physics we have to think holistically.

A man ceases to be a beginner in any given science and becomes a master in that science when he has learned that *this expected reversal is never going to happen* and that he is going to be a beginner all his life. RG Collingwood, *The New Leviathan*

01

Heat

DEFINITION:
HEAT: THE INTERNAL ENERGY POSSESSED BY ATOMS AND MOLECULES; THE TRANSFER OF SUCH ENERGY FROM BODIES WITH HIGHER ENERGY TO THOSE WITH LOWER ENERGY.

Human beings need to maintain a core temperature of 37.6°C for wellbeing (heat stroke sets in if the core temperature rises above 40.5°C and hypothermia if it drops to 35°C). There are physiological mechanisms that help regulate our body temperature (we sweat when we are too hot and shiver when we are too cold), as well as behavioural mechanisms (we stay out of the noonday sun). But there are also cultural adaptions: we wear clothes to keep us warm or cool; we light fires to keep us warm; and we construct buildings to provide shelter.

Until the middle of the 20th century, the main methods of regulating the internal temperatures of buildings were the use of individual heat sources (mainly open fires) and the use of air currents to introduce cooler air from outside. Now, however, heating and cooling technologies have improved, fuel is expensive and we have discovered the detrimental effects of the profligate use of energy to condition buildings in terms of pollution and climate change. The challenge for today and tomorrow is to maintain comfort conditions without an excessive use of energy (and, by extension, excessive greenhouse gas emissions).

In order to understand how to design buildings that provide a suitable internal temperature while minimising energy use, we need to understand how heat behaves in relation to the building fabric. This chapter begins by examining the fundamentals of heat, the effect of heat on materials, the mechanisms by which heat travels through them and the phenomenon of solar gain. It then considers heat transfer through the building fabric, looking at methods of measurement and the means by which heat transfer may be reduced, before considering thermal mass and how that may be measured.

The final section of the chapter addresses the energy performance of buildings, and examines how the control of heat transfer and the thermal mass of the fabric can contribute towards minimising energy consumption for heating and cooling.

Although this chapter does consider heat loss through the building fabric, we must remember that air movement can also result in significant heat transfer: this is discussed briefly in this chapter, but a more detailed treatment is given in chapter 2.

The fundamentals of heat

Let us begin by examining how materials respond to changes in temperature, and the mechanisms by which heat is transferred through the materials and air spaces that make up the building fabric. We can then consider how thermal insulation can reduce the heat transfer.

Heat and temperature

Heat is the internal energy possessed by matter: a material or object with a high level of internal energy has a higher temperature than a body with a low level of internal energy. There are two components of heat to consider:

* **Sensible heat** – the heat which results in a change in temperature of a material ('sensible' with the meaning of being sensed or felt)
* **Latent heat** – the heat involved in changing the phase of a material, without affecting its temperature (e.g. converting liquid water to water vapour requires energy, while solidifying a liquid wax will release energy)

Most of this chapter addresses sensible heat: a use of latent heat is also explored (see: Phase change materials p.35).

Temperature is the fundamental measure of heat, and is measured in kelvin, indicated by K. The scale starts at 0 K – absolute zero – with the triple point of water (the temperature and pressure at which water can be gas, liquid and solid at the same time) at 273.16 K. One kelvin is defined as 1/273.16 of the thermodynamic temperature of the triple point of water.

For day-to-day we use the Celsius scale, which has the melting point of ice (at standard pressure) at 0°C and the boiling point of water at 100°C. For conversion between the two scales, a kelvin is equivalent to a degree Celsius, so subtract 273.16 to convert kelvins to degrees Celsius, and add 273.16 to convert degrees Celsius to kelvin. It is worth noting that there are still a few countries (in particular, the United States) where the Fahrenheit scale is in use as part of the imperial system of weights and measures.

Determining the temperature experienced by building occupants requires the consideration of two different components of heat:

* The radiative temperature, resulting from radiation received and emitted (see: Radiation p.6);
* The air temperature.

Those two values are combined, by a weighting method which includes the effect of air movement, to give the **operative temperature** that can be used for design purposes.

Establishing a suitable design temperature can be even more difficult, because a building occupant's perception of whether a building is too hot or too cold can be affected by the prevailing outdoor temperature as well as factors such as dress codes.[1]

Heat capacity and thermal mass

The application of heat to a material raises its temperature, and the extent of that rise depends on the material: some require a lot of energy to warm them, others require less. The amount of energy needed to raise the temperature of a material is expressed by its **specific heat capacity**, which is the energy needed to raise the temperature of one kilogram of the material by one kelvin (measured in J/kgK, joules per kilogram kelvin). As table 1.1 shows, many construction materials have similar specific heat capacities.

The thermal mass of a material is also affected by its density, so a more practical measure is the **heat capacity,** which is the energy required to raise the temperature of a layer of a certain thickness. The heat capacity of a layer is calculated by multiplying its density, specific heat capacity and thickness (in metres). Table 1.1 shows that mineral wool and dense concrete blocks have a similar specific heat capacity, but very different densities, so the heat capacity of a 100 mm layer of concrete block is nearly a hundred times that of a 100 mm layer of mineral wool.

TABLE 1.1 HEAT CAPACITIES OF COMMON MATERIALS

MATERIAL	DENSITY (kg/m^3)	SPECIFIC HEAT CAPACITY (J/kgK)	HEAT CAPACITY OF A 100 mm THICK LAYER (kJ/m^2K)
Steel	7800	450	351
Dense concrete block	2300	1000	230
Brick	1750	1000	175
Dense plaster	1300	1000	130
Wood	500	1600	80
Autoclaved aerated concrete ('aircrete') block	700	1000	70
Plasterboard	700	1000	70
Polyisocyanurate (PIR) insulation board	30	1400	4.2
Mineral wool - batt	25	1030	2.6

SOURCE: DATA FROM BS EN ISO 10456 AND CIBSE GUIDE A3.

Heat transfer mechanisms

Where materials are hotter or colder than adjacent materials or surfaces there will be a transfer of heat from the hotter to the colder. There are three main mechanisms of heat transfer: conduction, mass transfer and radiation.

Conduction

Conduction heat transfer occurs as the molecules in solids, liquids and gases vibrate and collide against each other. Energy is passed from molecules with higher internal energy to those with lower internal energy, raising their temperature, which results in heat travelling through the material. Heat travels through different materials at different speeds: you can experience this when you touch a metal door handle and then the face of a wooden door. The metal handle feels colder, because it conducts heat away from your hand more quickly than the wood.

The rate of conduction heat transfer is affected by:

* The proximity of molecules – in solids the molecules are packed more tightly together than those in liquids, so energy can pass more easily between them. In liquids the molecules are further apart, so fewer collisions occur. In gases the distance between molecules is even greater, resulting in less frequent collisions and even slower energy transfer.
* The arrangement of the atoms (in solids) – for example, the atoms in metals are arranged in a tight lattice, which means the increased vibrations of the atoms produced by heating are transferred rapidly to neighbouring atoms.
* The presence in some materials of free electrons – if some electrons within a material are not attached to atoms (typically, in metals) the electrons move when electronic or magnetic fields are applied; they move easily when their energy state is raised, so transferring energy through the material.

The rate at which heat travels through a material is expressed in its **thermal conductivity**, which is the amount of heat transferred per second through one metre of a material, with a one kelvin temperature difference between one side and the other. The unit of conductivity is W/mK (watts per metre kelvin). The conductivity of a material is often referred to as its lambda value, λ.

Most construction materials are composites, consisting of many elements amalgamated in different proportions. In addition to solid matter, construction materials also contain matter in gaseous and liquid state (timber, for example, typically has a moisture content between 10% and 20%). Thermal conductivities are therefore established by laboratory testing of samples.

Table 1.2 sets out the conductivities of a number of common construction materials. The range of thermal conductivities is huge; metals have conductivities 425–8000 times greater than those of thermal insulants. To put those values into context: in order to achieve the same performance as 100 mm of expanded polystyrene insulation you would need 420 mm of autoclaved aerated concrete, 3.67 m of concrete and 56.7 m of stainless steel.

TABLE 1.2 THERMAL CONDUCTIVITIES FOR COMMON CONSTRUCTION MATERIALS	
MATERIAL	CONDUCTIVITY (W/mK)
Aluminium	160.0
Carbon steel	50.0
Stainless steel (austenitic)	17.0
Granite	2.8
Sandstone	2.3
Concrete (density 2400 kg/m³)	2.0
Concrete block (density 1800 kg/m³)	1.15
Glass	1.0
Brick (in exposed position)	0.77
Aircrete	0.20
Timber	0.13
Plasterboard	0.21
Thermal insulation*	0.019-0.040

SOURCE: DATA FROM BS EN ISO 10456
* Typical conductivities for different insulants are given in table 1.4.

Mass transfer

Heat is transferred in gases when molecules move by 'mass transfer' as a result of the mechanisms of **convection** and **advection**.

Convection occurs when part of a volume of gas is heated: the molecules gain more energy and random motion increases, making the heated gas less dense. The heated gas is then displaced by cooler, denser gas, making it rise. The continued presence of a heat source will result in currents of heated gas rising and cooler gas sinking, transferring energy with the molecules. You can sense convection in gases by holding your hands above a central heating radiator. Air in contact with the radiator is warmed (mainly by conduction) and expands. It is then displaced by cooler denser air and forced upwards, as shown in Figure 1–01.

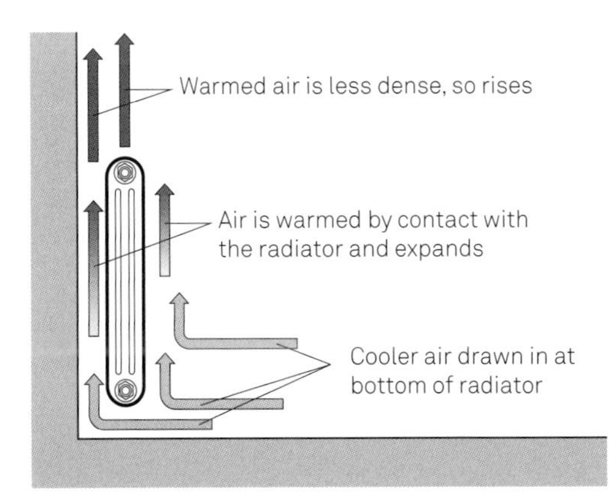

FIGURE 1-01
CONVECTION ENABLES HEAT FROM A RADIATOR TO WARM THE BUILDING INTERIOR

The rate of heat transferred by convection depends on the speed at which the gas moves (which is affected by its viscosity), the temperature difference, the direction of heat flow and the amount of energy required to raise the temperature of the gas (which is affected by its density and heat capacity).

Advection is the transfer of heat as the result of bulk movement of molecules in a gas driven by other forces, for example a difference in air pressure between the inside and outside of a building (see chapter 2 for a more detailed discussion of air movement).

Radiation

All bodies emit thermal energy in the form of **electromagnetic radiation** (which is discussed in more detail in chapter 5). Radiation can travel across a vacuum (e.g. radiation from the sun). A common experience of radiation is standing by a fire or stove in an otherwise cold room. The side of your body facing the fire is warmed by radiation from the fire, which is absorbed faster than you give off radiation. The other side of your body is colder because it emits more radiation to the environment than it receives.

The amount of radiation a body emits depends upon its temperature (radiation increases with temperature) and its **emissivity** (which expresses the radiation emission potential of the surface as a fraction of the potential of a perfect emitter – known as a black body). Emissivities lie between 0 and 1: a surface with a high emissivity will emit more radiation than one with a low emissivity. Most construction materials have high emissivities – above 0.8.

As well as emitting radiation, a body will receive radiation from other bodies, which will either be reflected or absorbed. The proportion of radiation absorbed at the surface of a body is measured by its **absorptivity**, which lies between 0 and 1: most common construction materials have high absorptivity and will absorb a high proportion of the radiation reaching them. A number of materials (e.g. glass) will transmit a proportion of the radiation reaching them. (This is discussed in more detail below: Controlling heat transfer through openings. p. 20)

The overall effect of radiation on the thermal energy of a body depends on the balance of radiation given out (emitted) and that received (absorbed). If all other heat flows are equal, a body which emits more radiation than it absorbs will cool, while one which absorbs more than it emits will be warmed.

Heat transfer through solids

The thermal performance of a layer of a material in a construction is expressed by its **thermal resistance**, which is the resistance to heat transfer of a specific thickness of a material. Thermal resistance is expressed in m^2K/W (metre squared kelvin per watt): a higher thermal resistance is better for reducing the rate of heat transfer.

For solid materials, where conduction is the main mechanism of heat transfer, the thermal resistance is calculated by dividing the thickness of the material (in metres) by its thermal conductivity. The relationship between thickness and thermal resistance is straightforward: doubling the thickness of a material doubles its thermal resistance (see Figure 1–02). For the same thickness, a material with low thermal conductivity will give a higher resistance than a material with high thermal conductivity.

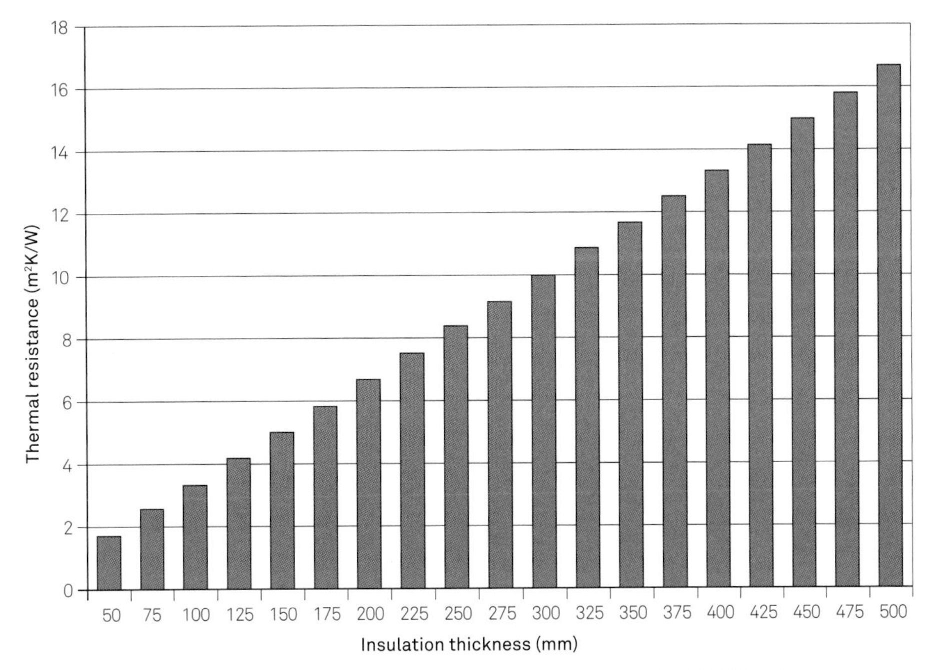

FIGURE 1-02 THE RELATIONSHIP BETWEEN THERMAL RESISTANCE AND THICKNESS

Heat transfer through air spaces

The thermal performance of air spaces within building elements is affected by all three heat transfer mechanisms.

Conduction in air spaces

In narrow air spaces conduction accounts for more heat transfer than convection, but beyond certain thicknesses (approximately 12 mm for heat transfer upward, 20 mm for horizontal heat transfer and about 360 mm for downward heat transfer) substantially more heat transfer occurs by convection.

Convection in air spaces

Temperature differences across air spaces will generate convection currents. Air in contact with the warmest face of the air space will be heated, and will expand and be forced upward by colder, denser air sinking to replace it. Because convection currents move warmed air upward (together with the energy it contains) the effect of heat transfer across an air space varies with the direction of heat transfer through the whole element:

* Where heat transfer is predominantly upward, convection works in the same direction and will increase heat transfer. This occurs in roof cavities when heat is being transferred out of the building and also in floor cavities when heat is being transferred into the building.
* Where heat transfer is predominantly horizontal, convection works at right angles to that transfer, and will make a small increase to the rate of heat transfer. This occurs in wall cavities.
* Where heat transfer is downward, convection works in the opposite direction, reducing the rate of heat transfer. This occurs in floor cavities when heat is being transferred out of the building, and in roofs, when heat is being transferred into the building.

Any air currents within an air space – whether deliberately induced (see chapter 3: Ventilation for the fabric p. 96) or as a result of air **infiltration** – will increase heat transfer through the airspace. (A positive effect of a convection current in a wall cavity is described below: Beneficial air movement in air spaces.) If the air space is connected to outside air the thermal performance of the entire construction can be severely compromised (see: Heat loss in party walls p. 24).

Radiation in airspaces

Every surface bounding an air space in an element emits and absorbs radiation. If one surface of the air space emits more radiation than it receives there will be a net transfer of heat across the air space. Because warmer surfaces emit more radiation than cooler ones, the direction of heat transfer will generally be from the warmer side of the air space to the cooler side. The rate of radiation heat transfer across an air space depends on the temperature difference across the space, and the emissivities of its surfaces.

Determining thermal resistances of air space

There is no simple relationship between the thermal resistance of an air space and its thickness because heat transfer across an air space involves all three mechanisms of heat transfer. Figure 1–03 shows the resistances of air spaces for upward, horizontal and downward heat flows. In each case the resistance increases with thickness, but reaches a plateau when the effect of convection heat transfer exceeds that of conduction heat transfer.

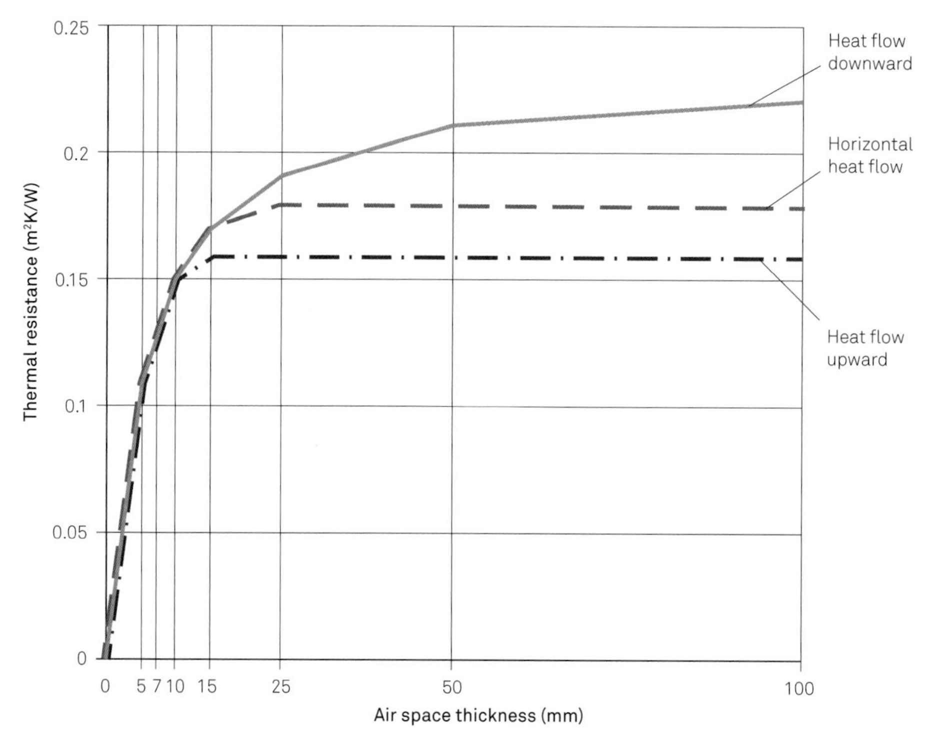

FIGURE 1-03 AIR SPACE RESISTANCE BY THICKNESS

Heat transfer at surfaces of building elements

The overall rate of heat transfer through building elements is affected by the conditions at the interior and exterior surfaces of the elements. At interior surfaces, temperature-driven convection currents account for much of the heat transfer: the rate of transfer being greatest when the direction of transfer is up, rather than down. At exterior surfaces air movement produced by the wind is more significant than temperature-driven convection; in these circumstances, the large amount of air movement occurring at the surface means that conduction heat transfer has a negligible effect.

Radiation heat transfer also occurs at interior and exterior surfaces: the rate of transfer is determined by the balance of radiation emitted and received by the surface.

The thermal performances of interior and exterior surfaces are expressed as surface resistances which combine the effects of all mechanisms of heat transfer. Commonly used values are set out in table 1.3.

TABLE 1.3 DEFAULT SURFACE RESISTANCES (m^2K/W)

	DIRECTION OF HEAT FLOW		
	UPWARD	HORIZONTAL	DOWNWARD
Internal surface resistance	0.10	0.13	0.17
External surface resistance	0.04	0.04	0.04

SOURCE: DATA FROM BS EN ISO 6946:2007.

Radiation heat loss can also result in a substantial lowering of the external temperature of roofs at night. The rate of radiation heat transfer from a roof is the balance of radiation emitted from the surface and that absorbed from the sky and clouds. On clear, still nights the rate of radiation from the roof far exceeds the incoming radiation, resulting in rapid heat loss from the roof surface (commonly referred to as 'night time radiative cooling'), which can be beneficial in a climate where cooling is required.

On well-insulated roofs, the rate of loss is far higher than the movement of heat from the building interior, resulting in a surface temperature substantially lower than the air temperature, which is why night-time radiative cooling can result in transient condensation underneath the roof covering (see chapter 3: External climate p. 92).

Heat and light

Light from the sun, including infrared, visible and ultraviolet light, is a form of radiation; when it reaches the surface of a building it can be reflected, absorbed and transmitted. Opaque materials will reflect or absorb radiation; as a material absorbs radiation it gains energy and its temperature will rise, the extent of that warming depending on the wavelength and intensity of the light, the absorptivity of the surface and the heat capacity of the material. As the surface warms, heat will be transferred by conduction towards the building interior.

Translucent materials, such as glass, reflect and absorb radiation, but they also transmit some infrared and visible light to the building interior. The amount of solar radiation which is transmitted depends on the characteristics of the glazing (see: Heat transfer through openings p. 16). The light will fall on the interior surfaces of the building, where a large part of its energy will be absorbed by the fabric, raising its temperature. Some of the energy will be re-radiated at longer wavelengths, transmitting the heat around the interior of the building, while some will be lost to the air inside the building. The overall effect of this **solar gain** will be to raise the building's internal temperature.

Thermal insulation

Thermal insulation is used to reduce heat transfer through floors, walls and roofs. There are many different types of insulation, but virtually all work by trapping small pockets of low conductivity gas within their bulk. (The exception is the vacuum panel, consisting of microporous core which is evacuated then sealed in a gas-tight cover.) The main types of insulation are:

* Closed-cell insulants – formed with bubbles of gas in a rigid matrix that may be made of plastic in the case of extruded polystyrene (XPS), polyurethane (PU), polyisocyanurate (PIR) and phenolic insulation, or glass in the case of foamed glass insulation. Aerogel – a very light, translucent solid – also consists of gas trapped within a rigid matrix.
* Expanded bead insulation – typically in the form of expanded polystyrene (EPS), which is available bonded as boards or as loose beads.
* Fibrous insulation – in which pockets of still air are formed by strands of material, which may be mineral fibre, glass fibre, cellulose (usually from recycled paper) or 'natural' materials such as sheep's wool.

The thermal performance of an insulant depends on how well it addresses the three mechanisms of heat transfer:

* Conduction through the solid material – the rate of conduction depends on the type of material and its density, with lower densities reducing the amount of material through which heat can be conducted.
* Conduction through the gas – the conductivity of the gas is key to overall performance, with air having a conductivity of 0.025 W/mK, argon 0.016 W/mK and pentane 0.014 W/mK.
* Convection – for closed-cell insulants, convection is insignificant, because the isolation of the cells precludes convection currents. However, convection is an issue in fibrous insulants, because differential temperatures across the insulation generates convection currents. Where the direction of heat transfer is upward (in a heat-out roof or heat-in floor), convection currents will increase the rate of transfer.
* Radiation – there is some radiative heat transfer across the air spaces within insulation, with the rate of transfer dependent on the surface characteristics of the cells or fibres. The only insulants which specifically address radiative transfer are graphite-enhanced polystyrene insulation (EPS and XPS), which have lower conductivities as a result of the graphite granules reducing radiative transfer.

Table 1.4 gives typical thermal conductivities for common insulants.

TABLE 1.4 TYPICAL CONDUCTIVITIES OF THERMAL INSULATION	
MATERIAL	CONDUCTIVITY (W/mK)
Mineral fibre batt	0.038
Loose fill mineral fibre	0.040
Expanded polystyrene (EPS)	0.036
Expand polystyrene with graphite granules	0.030
Extruded polystyrene (XPS)	0.027
Polyisocyanurate (PIR)	0.022
Phenolic insulation	0.019
Aerogel	0.013

To ensure that the values realistically represent the performance of the fabric, the stated thermal conductivity of an insulant should account for:

* Variations in production – The performance of insulation will inevitably vary within and between production batches. Plotting the measured conductivity from different batches will typically result in a bell curve (Figure 1–04). Using typical, or mean, values of conductivity results in half the samples having a better conductivity than that quoted, and half the samples being worse. To avoid overestimating installed performance European thermal insulation standards require thermal conductivities to be quoted as lambda 90/90 ($\lambda_{90/90}$) values, which means 90% of samples have thermal conductivity equal to or better than the value quoted, with a 90% level of certainty.
* Long-term change in performance – The thermal conductivity of closed-cell plastic insulation materials (such as PIR, PU and XPS) will change over time, initially as oxygen and nitrogen in the atmosphere migrate into the insulation and change the balance of gases within the cells, and later as the blowing agent migrates to atmosphere. (One reason foil facings are used on PIR boards is to reduce gas migration and improve long-term performance.) The performance of fibrous insulation will also change, for example as a result of deterioration of the chemicals which bind the fibres together. Calculations of long-term performance need to use the thermal conductivities which reflect the deterioration of performance over time: these are commonly referred to as 'aged' conductivities.
* Variations with temperature – The thermal conductivity of insulation increases with temperature, mainly as a result of the higher conductivity of the blowing agent or air in the insulant. Fibrous insulants will also experience increased internal convection, further increasing the thermal conductivity. Where the in-use temperature of insulation is significantly different from the temperature at which the thermal conductivity was obtained, the conductivity or thermal resistance should be adjusted for temperature.[2]
* Moisture – Water is a substantially better conductor of heat than air or other gases (thermal conductivity 0.60 W/mK, compared with 0.025 W/mK for air and 0.014 W/mK for carbon dioxide). Consequently any water absorbed by an insulant will increase its thermal conductivity (i.e. reduce its capabilities as an insulator). Where insulation is to be installed in damp conditions – for example in foundations – the potential for water absorption should be taken into account when calculating the amount/type of insulation material needed.

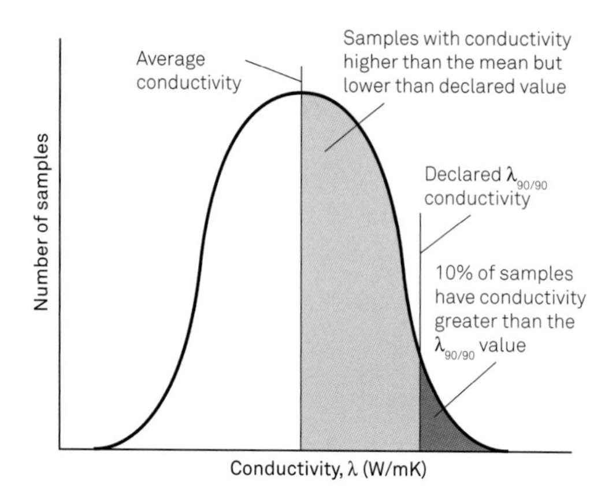

FIGURE 1-04
DETERMINING THE QUOTED THERMAL CONDUCTIVITY OF INSULATION

Thermal bridging

Many constructions contain a layer of material which is regularly interrupted by a second material, for example, insulation interrupted by the timber studs, or mortar joints in aircrete blockwork: this is referred to as **thermal bridging** (sometimes incorrectly termed 'cold bridging'). The interrupting material generally has a higher thermal conductivity than the main material, so the rate of heat transfer will be higher than if there were no bridging, and the U-value (see: Heat transfer through building elements p. 13) will be higher. (An exception to this is mortar bridging in stone walls, where the mortar has a lower conductivity that the stone) Figure 1–05 shows the effect on the temperatures through a wall where a layer of insulation is bridged by a structural support.

As well as affecting the rate of heat transfer, thermal bridging can also affect surface temperature, increasing the risk of surface condensation (see chapter 3: Controlling surface condensation p. 84).

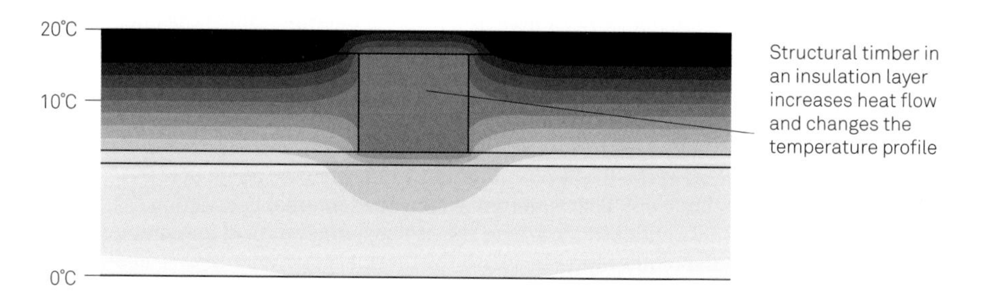

FIGURE 1-05 TEMPERATURE CHANGES CAUSED BY THERMAL BRIDGING

Assessing heat transfer through the building fabric

We have to be able to quantify heat transfer so that we can assess the overall energy performance of buildings. The same principles apply for measuring heat transfer from the building to its surroundings (the northern European winter), or to the building from its surroundings (the Australian summer).

Heat transfer through building elements

The rate of heat transfer through a building element is dependent on the performance of the materials and air spaces and surfaces of which it is composed. In the UK and Europe the standard measure of the thermal performance of a building element is the **thermal transmittance**, commonly referred to as the **U-value**. The U-value is the rate of heat transfer through one square metre of an element's surface, with a temperature difference of one kelvin between one side and the other. The units are W/m^2K (watts per metre squared kelvin). Under identical conditions, an element with a low U-value will transmit less heat than an element with a high U-value.

The U-value of an element is the inverse of the total thermal resistance ($U = 1/R$).[3] Figure 1–06 shows the effect of increasing the thickness of thermal insulation in a construction: the resistance increases steadily, while the U-value drops rapidly to start with (adding 50 mm of insulation to an existing 50 mm improves the U-value from 0.6 W/m^2K to 0.3 W/m^2K), until it reaches a point where additional insulation makes very little difference (adding 50 mm to an existing 250 mm only improves the U-value from 0.12 W/m^2K to 0.10 W/m^2K). When looking to improve the thermal performance of a building's fabric it is important to compare the benefit which would be achieved by adding the same amount of insulation to different elements.

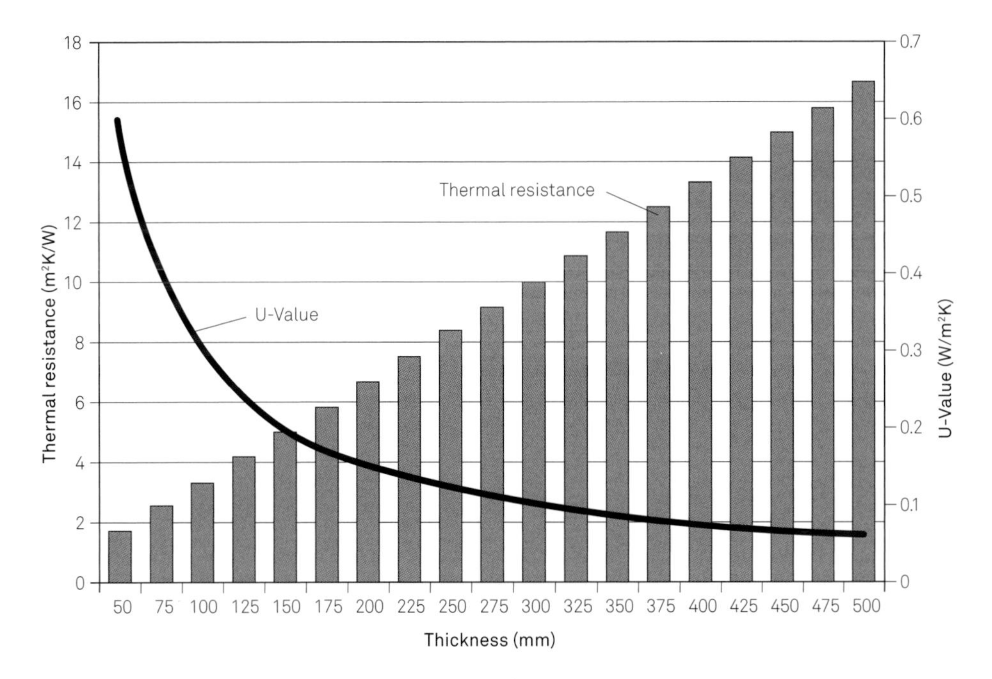

FIGURE 1-06 THERMAL RESISTANCE AND U-VALUES FOR INCREASING THICKNESSES OF INSULATION

Direct measurement of thermal transmittances

U-values may be determined from laboratory or field measurements. Typically, laboratory measurements are carried out using 'guarded hot box' techniques. Essentially, part of an element, or a component such as a window, is placed inside a highly insulated box. The spaces to each side of the sample are maintained at different temperatures: one high, one low. The amount of energy needed to maintain the high temperature is measured, allowing the rate of heat loss to be measured, and the U-value to be calculated.

The U-values of existing walls can be calculated using heat flux sensors on either side of the wall. In situ measurements are particularly suitable for establishing the performance of existing walls of unknown construction.

Calculation of thermal transmittance

When an element has no thermal bridging the U-value can be calculated from the total resistance. However, most elements have thermally bridged layers so instead we use the **combined method**, as defined in ISO 6946[4] and described below (See box: The combined method p. 15).

Complex elements, such as those containing steelwork or irregularly shaped components are beyond the scope of the combined method and must be analysed by computer-based **numerical modelling**, which uses a two or three-dimensional computer model of an element.

The thermal analysis software imposes a network of points (nodes) throughout the model, then estimates the heat flows across the nodes to produce an initial value for heat transfer. That approximate solution is refined by an iterative process which recalculates the heat flows many times, finishing when recalculations make no significant difference to the results.

Numerical modelling takes longer than the combined method calculations, but is more accurate. It is also used to determine heat loss at junctions and to assess the effects of fixings and brackets on insulation layers.

Neither the combined method, nor numerical analysis, account for the effect of air infiltration between outside and inside through the building fabric, although the combined method can make an approximation of the effect of ventilated cavities. Heat loss through air infiltration is usually considered for the whole building (see: Whole building energy performance and chapter 2).

Heat transfer through the ground

Heat transfer through elements in contact with the ground – ground floors and basement walls and floors – is affected by the conductivity and heat capacity of the ground and the ground temperature. The temperature varies on an annual cycle, and also varies across the floor; in heat loss conditions, the ground beneath a floor will be warmer at the centre than at the perimeter (see Figure 1–07).

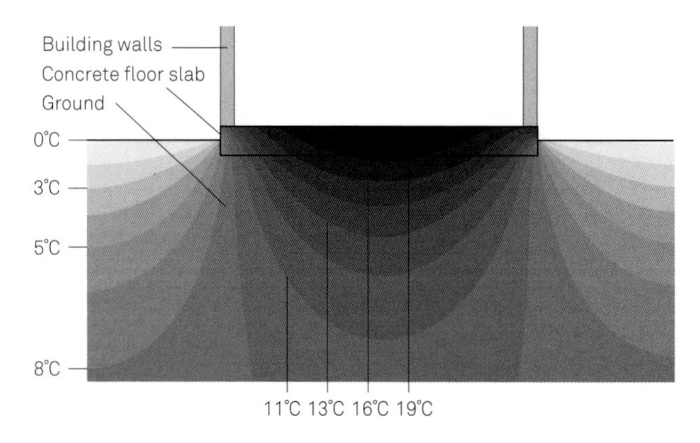

FIGURE 1-07
TEMPERATURE PROFILE BELOW A GROUND-BEARING SLAB

THE COMBINED METHOD

The combined method calculates U-values using (a) the resistances of the layers in the construction and (b), for each thermally bridged layer, the fractional areas that represent the proportions of each material in the layer.

The calculation proceeds by determining the thermal resistance of the construction in two ways (R_{max} and R_{min}), as shown in the following diagrams.

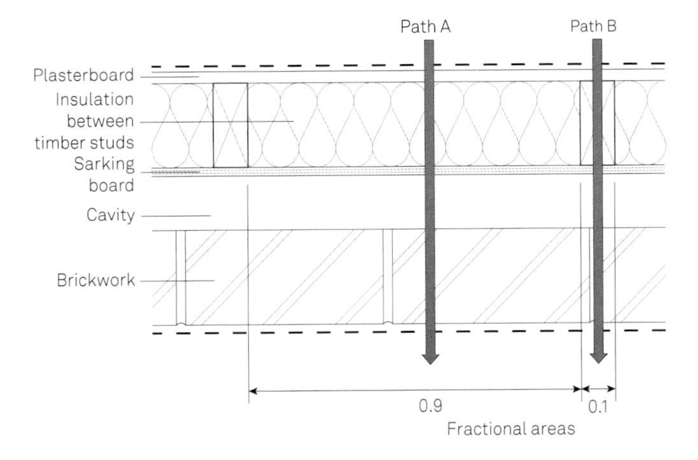

R_{max} is the total resistance calculated from heat paths through the construction (referred to in standards as the 'upper limit of thermal resistance'). The calculation involves the resistance and fractional area of each heat path, as illustrated here for a plan section through a timber-framed wall, showing two heat paths and their fractional areas.

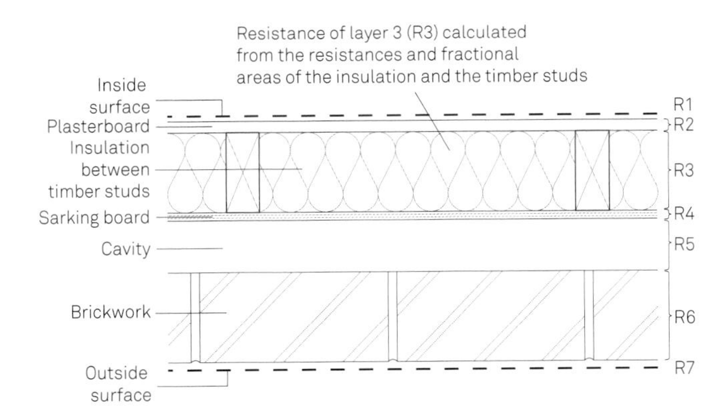

R_{min} is the total resistance calculated from the resistances of individual layers (referred to in standards as the 'lower limit of thermal resistance'), taking account of bridging in the layer, as illustrated in this section of timber-framed wall.

In any construction containing thermally bridged layers R_{max} will (as the names suggest) be bigger than R_{min}: the actual total resistance lies somewhere between the two and, for the purposes of the calculation, is taken to be their average. The U-value is then the reciprocal of the total resistance.

The full combined method includes adjustments for mechanical fasteners and air leakage, as well as modifications for tapered insulation layers and inverted roofs.

Although the overall rate of heat loss through a large floor will be greater than that through a smaller floor of the same construction, the rate of heat loss per square metre will be less for the larger floor, giving a lower average U-value. A modified version of the combined method calculation takes account of these effects by considering the ratio between the floor's perimeter and its surface area: it also allows for heat loss through the sub-floor void in suspended ground floors.

Heat transfer through openings

The thermal performance of glazing in openings (doors, windows or rooflights) and curtain wall systems depends on the performance of the glazing and the framing. The 'glazing' of a building may, of course, not be glass, but polycarbonate, aerogel or some hitherto undiscovered substance. For the sake of simplicity I will refer to it as glass.

The glazing is subject to conduction, convection in gas-filled cavities and radiation heat transfer in both directions. (The effect of radiation heat transfer in the form of solar gain is discussed below: Controlling heat transfer through the fabric. p. 17) Glass itself has a high thermal conductivity (1.0 W/mK, comparable with that of concrete blockwork), so most of the thermal performance in a double- or triple-glazed unit comes from the air spaces between the panes of glass. Cavity widths are limited to prevent convection.

The thermal performance of the glazing may be given as a **centre pane U-value**, which is the U-value measured perpendicularly through the centre of the glazing, taking no account of the effects of glazing spacers and edge seals.

Heat transfer within the frame takes place by conduction and, in hollow frames, by convection. Convection transfer in hollow frames is addressed by dividing the frame into smaller chambers which may sometimes be filled with foam insulation.

The overall U-value for an opening takes account of the performance of the glazing, spacers and frame. The U-value will vary according to the relative proportions of glazing and framing so the performance is often quoted for a standard window size and configuration.

The U-value of a window only addresses conduction heat transfer, so alternative measures, such as the Window Energy Rating (and the related Door Energy Rating), have been developed to express overall performance, taking account of conduction heat transfer, air leakage and the beneficial effects of solar gain. The Window Energy Rating is expressed in A–G bands, with A being the most energy efficient.

Heat transfer at junctions

A significant amount of heat transfer occurs at the junctions between building elements (e.g. the wall–floor junction) and at the edges of openings (e.g. the window jamb). Heat transfer through such junctions is expressed by the **linear thermal transmittance** or ψ-value (psi-value), which is the rate of heat transfer through one metre length of the junction, with a temperature difference of one kelvin between one side and the other. (The effect that the additional heat loss has on the temperature at the junction is calculated using the 'temperature factor', which is discussed below: chapter 3, Box: Temperature factor p. 88) The units are W/mK (watts per metre kelvin).

Calculating the linear thermal transmittance for a junction involves numerical modelling. For many junctions it is sufficient to model in a two-dimensional cross-section, but for complex constructions, particularly those which include steel framing, three-dimensional modelling is required.

Controlling heat transfer through the fabric

Having established how the rate of heat transfer can be determined we can now consider how to control heat transfer through the various parts of the building fabric in order to minimise the energy required to maintain comfort conditions. This section considers improvements to the thermal performance of building elements, the reduction of heat loss at junctions and the effect of air infiltration, as well as examining the performance of glazing.

The selection of thermal insulation

The use of thermal insulation is essential for reducing heat transfer through the elements that separate the conditioned interior of a building from outside. When selecting an insulant for a particular application there are six main criteria to consider:

* Thermal performance
* Thickness
* Moisture
* Strength
* Air movement
* Installation

Other considerations – beyond the scope of this book – might include embodied carbon, a requirement for local sourcing and price. Suffice to say that the better an insulant performs in the aspects listed above, the more it will cost.

Thickness

Any insulant could provide a specified U-value, but the required thickness would vary with its thermal conductivity. In some constructions – such as a cold pitched roof – there are no constraints on thickness, which means that an insulant with a higher thermal conductivity (say, 0.040 W/mK) may be used. In other constructions – notably walls – space constraints require the use of insulation with a lower conductivity (say, 0.022 W/mK) to provide the required thermal performance.

Moisture

The presence of water in an insulant will affect its thermal performance. The extent of deterioration depends on:

* The extent of exposure to moisture
* The rate at which the material takes up water
* The effect of moisture on the thermal performance

Closed-cell insulants are the least susceptible to moisture, because their structure makes it difficult for water to travel through the material by capillary action (see chapter 3: Capillary action p. 78). Fibrous insulants have an open structure and are more susceptible to moisture, although moisture-repellent treatments applied during manufacture will reduce water transmission. Nonetheless, 1% water content (by volume) can more than double the thermal conductivity of fibrous insulation.

Where long-term exposure to moisture is expected (e.g. below ground or on inverted roofs), a closed-cell insulant will be required.

Strength

The physical performance of insulation under a load is expressed as its **compressive strength**, which is the pressure required to compress the material by 10% or to break it. However, long-term performance is measured by constant compressive stress, typically to give 2% deformation over 50 years. Constant compressive stress values will be several times smaller than the compressive strength.

In many applications there is very little loading on insulation, but floor insulation will be subject to the dead loads of the floor construction and the live loads resulting from the occupation of the building. Floor loads in domestic and low-rise commercial projects can usually be met by mineral fibre and closed-cell insulants, but specialist applications, such as cold storage areas, require very high compressive strengths, with XPS being the only realistic option.

Air movement

The structure of fibrous insulation allows the air within it to circulate as a result of convection currents driven by the warming of air at the bottom of the insulation, increasing the rate of heat transfer through it. Air currents across the surface of fibrous insulation will draw out warmed air, replacing it with colder air.

Installation

For insulation to achieve anything close to the design performance it must be installed properly, forming a layer without gaps, fitted tightly against adjoining parts of the construction. Where insulation is to be fitted in a continuous layer (e.g. over a floor or flat roof, or across the face of a wall) rigid boards and batts are suitable. The use of interlocking boards with tongued and grooved or ship-lapped edges will avoid gaps and reduce air infiltration at board junctions.

Where insulation is to be fitted between structural members it is easier to get a tight fit with flexible material, such as mineral fibre, rather that trying to cut rigid boards to size. This is especially true for refurbishment projects where insulation is to be fitted around irregularly shaped and spaced timbers.

Reducing the effect of thermal bridging

The effect of thermal bridging on an element's thermal performance depends on:

* The relative conductivities of the two materials in the bridged layer – where the conductivities are close (e.g. mortar (0.94 W/mK) and brick (0.77 W/mK)) bridging will have a negligible effect, but the effect will be substantial where insulation layers (typically 0.022–0.040 W/mK) are bridged by timber (0.13 W/mK) or steel (15 W/mK);
* The relative proportions of the two materials – Figure 1–08 shows how the U-value of a timber-framed wall changes if the proportion of timber bridging the insulation layer is increased (note that 15% bridging is the default percentage for such calculations).

Ideally, thermal bridging should be avoided, but there are many constructions, such as those with timber and steel framing, in which space constraints make it inevitable. In such cases the effects of thermal bridging may be mitigated by:

* Increasing the thickness of the bridged insulation layers
* Reducing the proportion of bridging, (e.g. large-format aircrete blocks with thin adhesive joints give 1.3% bridging compared with 6.7% for standard-format blocks laid in mortar)
* Replacing some insulation between framing with a continuous layer of insulation over the face of the framing
* Using proprietary components to form thermal breaks and reduce heat transfer: this is particularly important where steel brackets and fasteners penetrate insulation layers

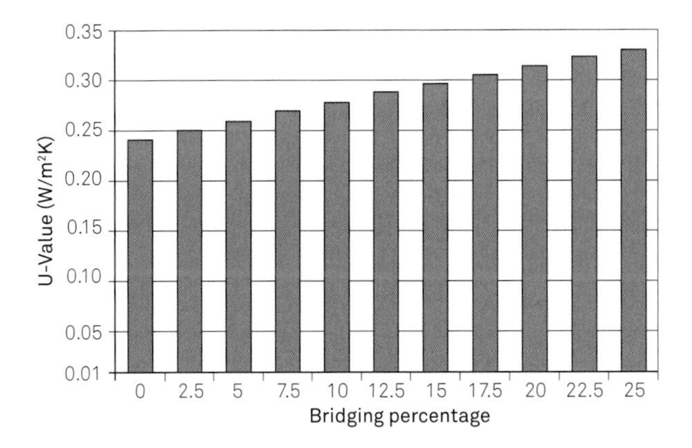

FIGURE 1-08
THE EFFECT OF
BRIDGING PERCENTAGE
ON THERMAL PERFORMANCE

Improving the thermal performance of air spaces

The rate of radiation heat transfer across an air space depends in part on the emissivity of its surfaces, so it is possible to reduce the rate of transfer by forming one or more of the surfaces of the air space from materials with low surface emissivities (typically 0.03–0.05) such as:

* Reflective metal facings of PIR insulation boards
* Membranes with metalised surfaces
* Encapsulated aluminium foil membranes
* Multi-foil membranes

Table 1.5 shows the increase in thermal resistance of a 25-mm wide cavity when one or both surfaces have low emissivity. As with any cavity, the direction of heat transfer affects the thermal resistance. When assessing the long-term performance it is important to use an emissivity value which takes account of the effects of dust and contamination of the surface and the consequent reduction in thermal resistance. Note also that low-emissivity surfaces are also reflective: in some situations an anti-glare coating is required to reduce the risk to installers (and the coating will also increase the emissivity).

Although low-emissivity surfaces do improve the thermal resistance of an air space, in virtually all cases thermal insulation to the same thickness would provide an even higher resistance. It is therefore sensible to use low-emissivity surfaces only where there are designed air spaces in the construction, rather than deliberately introducing an air space.

TABLE 1.5 CAVITY RESISTANCES FOR DIFFERENT SURFACE EMISSIVITIES (m^2K/W)

DIRECTION OF HEAT TRANSFER	SURFACE CHARACTERISTICS		
	TWO STANDARD SURFACES	ONE LOW-EMISSIVITY; ONE STANDARD SURFACE	TWO LOW-EMISSIVITY SURFACES
Vertical, heat-up	0.163	0.454	0.481
Vertical, heat-down	0.193	0.798	0.884
Horizontal	0.184	0.665	0.724

Note: These figures are calculated to BS EN ISO 6946:2007 for a 25 mm deep unvented airspace, with a mean temperature of 10°C. Standard surface: emissivity = 0.9; low-emissivity surface: emissivity = 0.05.

Controlling heat transfer through openings

When we consider heat transfer through openings we must first address heat transfer which occurs by a combination of conduction, convection and radiation, and then consider the transfer of solar radiation and light.

Heat loss

The rate of heat transfer through openings is typically six to eight times greater than that of the elements in which they sit. This is hardly surprising, given that the thermal conductivity of glass is similar to that of concrete and it is installed in thin layers.

Heat transfer can be reduced by using double or triple glazing in which two or three panes of glass are separated by narrow air spaces. (Although wider cavities would reduce conduction heat transfer, they would simultaneously increase convection heat transfer.) Further reductions in heat transfer may be made by filling the cavity with a gas that has a lower conductivity than air, such as argon and krypton.

The effect of glass on radiation heat transfer depends on the wavelength of the radiation (see chapter 5: Light and materials p. 21). Visible light and infrared light are transmitted, but radiation with longer wavelengths, such as that emitted by the internal surfaces of the building, is largely absorbed by the glass.

Applying a low-emissivity coating on one of the internal surfaces of double or triple glazing can reduce the amount of radiation transmitted across the cavity (in the same way as a low-emissivity surface does in a wall cavity).

There are two main types of low-emissivity coating:

* Soft coat – multiple thin layers of silver applied after manufacture of the glass. Soft coatings give emittances of 0.05–0.10, so are more effective than hard coatings, but they are susceptible to moisture and physical damage.
* Hard coat – modified tin oxide coatings applied as part of the manufacturing procedure. Hard coatings give emittances of 0.10–0.20. They are more resistant to damage than soft coat, but are also more expensive.

The thermal performance of existing windows can be improved by the application of low-emissivity window films. Table 1.6 summarises the typical range of U-values which can be obtained with different glazing configurations.

TABLE 1.6 TYPICAL THERMAL PERFORMANCE OF WINDOWS (WOOD FRAME, ALL UNITS WITH 12 MM GAP)	
GLAZING CONFIGURATION	U-VALUE (W/m²K)
Single-glazed	4.8
Double-glazed, air-filled	2.8
Double-glazed, argon-filled	2.7
Double-glazed, air-filled, low-e, hard coat	2.2
Triple-glazed, air-filled	2.1
Double-glazed, air-filled, low-e, soft coat	2.0
Triple-glazed, argon-filled	2.0
Double-glazed, argon-filled, low-e hard coat	2.0
Double-glazed, argon-filled, low-e soft coat	1.8
Triple-glazed, air-filled, low-e hard coat	1.7
Triple-glazed, air-filled, low-e, soft coat	1.6
Triple-glazed, argon-filled, low-e, hard coat	1.6
Triple-glazed, argon-filled, low-e, soft coat	1.4

SOURCE: SAP 2012 TABLE 6E.

Of course, other components contribute to the overall thermal performance, including the spacers of double- or triple-glazed units and the framing members. Metal frames present a particular challenge, given the high thermal conductivity of steel and aluminium: the frames should be formed with thermal breaks, typically of neoprene rubber, to isolate the interior components from the exterior components and so reduce conduction heat transfer.

Solar transmission

When radiation from the sun reaches the glazing of an opening some radiation will be reflected back from the glass, some will be absorbed and re-radiated at a longer wavelength, and some will be transmitted to the building, where it will provide illumination and warm the interior.

The fraction of solar radiation which passes through glazing is expressed by the g-value, which is a value between 0 and 1. In some countries the solar heat gain coefficient (SHGC) is used. This is effectively a g-value for the whole window, taking account of the glazing, framing and screens.

The g-value may be quoted for the glazing of an opening or for the whole component (this is sometimes indicated with a 'window' subscript: g_{window}). Typically, standard double glazing will have a g-value in the range 0.60–0.75, while triple glazing will have lower g-values.

The proportion of visible light passing through a window can be expressed as the light transmittance. The light transmittance value for a window will be similar but not identical to its g-value, because glazing responds differently to visible and infrared wavelengths.

Where solar gain will be beneficial to reduce energy use, glazing with high g-values is preferable. However, where solar gain is likely to lead to overheating, solar control glazing should be used to limit heat gain. This may involve tinted glass or reflective coatings, which will also reduce the transmission of visible light.

A better solution is to use glass with spectrally selective coatings, which will transmit a high proportion of solar radiation at visible wavelengths, but a much smaller proportion of infrared radiation, so maintaining the benefit of daylight, but reducing heat gain. Low-emissivity (low-e) coatings intended to reduce heat transfer will also reduce the amount of solar radiation passing through the glazing and so limit solar gain.

Reducing heat transfer at junctions

The reduction in elemental U-values over recent decades has resulted in heat-transfer at junctions forming a much greater proportion of overall heat transfer. For example, in a dwelling built to meet the regulatory standards of 1995, junction heat loss would account for roughly one-fifth of fabric losses, but junction heat loss would account for two-fifths of losses in the same dwelling built to meet 2013 standards. Consequently, any serious attempt to reduce heat transfer must attend to junctions.

There are two main constraints which affect junction heat loss:

* The geometry of some junctions results in a greater exposed heat loss surface on the exterior than on the interior (Figure 1–09), resulting in additional heat loss. For such junctions it is therefore not possible to eliminate the additional heat loss, only reduce it.
* Structural requirements often prevent the formation of a continuous layer of insulation across the junctions. For example, Figure 1–10 shows a conventional wall junction, where the loadbearing structure of the wall prevents the insulation in the ceiling from connecting with the insulation in the wall.

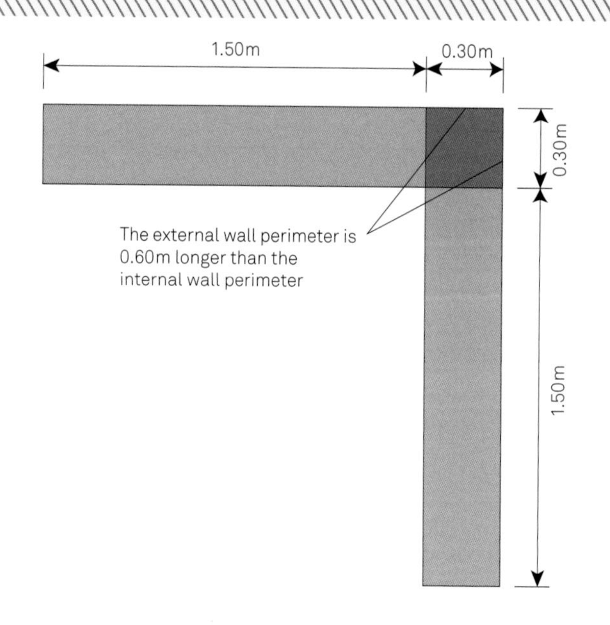

FIGURE 1-09
PLAN SECTION OF A WALL JUNCTION SHOWING THE GREATER EXTERNAL SURFACE AREA

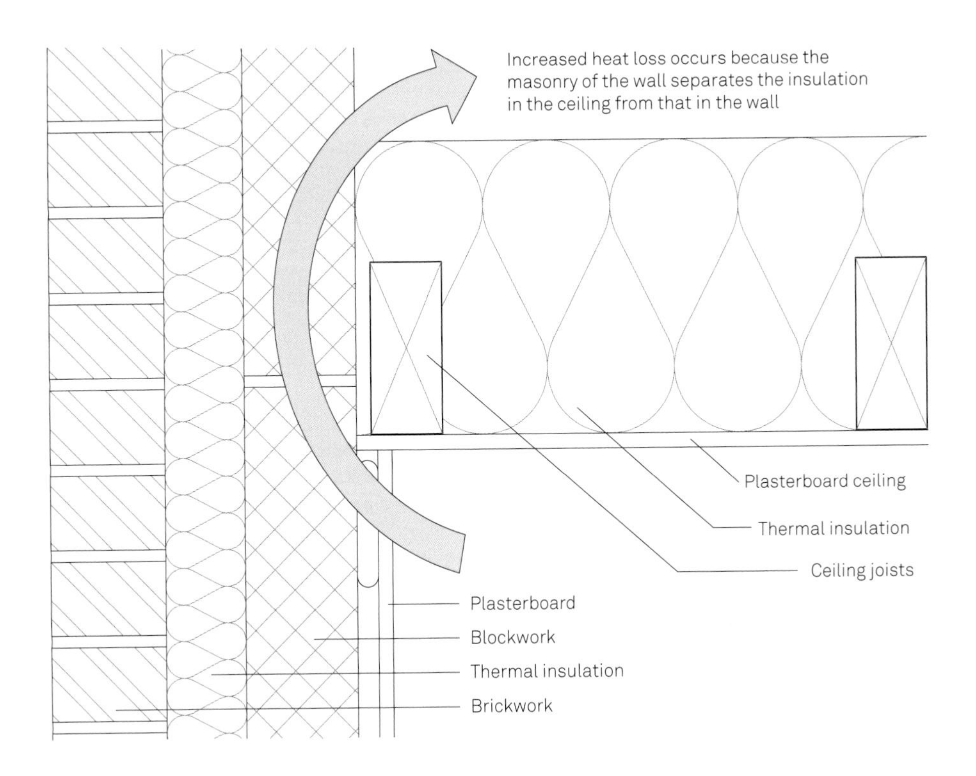

FIGURE 1-10 CROSS-SECTION OF THE WALL-CEILING JUNCTION AT A GABLE

Within those constraints, it is possible to minimise the heat loss at junctions by following four rules of thumb:

* Minimise the number of junctions – the use of several different construction types to form the thermal envelope will produce more junctions, more discontinuities and more heat loss.
* Set door and window frames to overlap the insulation plane, preferably rebating them into the full thickness of the insulation.
* Specify lintels carefully. Steel lintels transfer heat rapidly through walls because of their high thermal conductivity. Use well-insulated lintels and consider using two separate lintels where the wall contains a cavity.
* Overlap insulating layers to reduce bridging paths. This is particularly useful to address wall/floor junctions and wall/roof gable junctions (see Figure 1–10) where the structural constraints prevent continuity of insulation.

Good thermal detailing will also produce a more airtight structure and so contribute to reducing heat transfer by air leakage (see chapter 2: Infiltration p. 48).

Air infiltration

Air movement by ventilation and by infiltration will also move heat into and out of the building. As will be seen in chapter 2 it is important to maintain good internal air quality: the best strategy is to eliminate unintended air movement and deliberately supply the necessary clean air, adopting a strategy of 'build tight, ventilate right'.

HEAT LOSS IN PARTY WALLS

In the UK a common method of preventing sound transmission between semi-detached or terraced houses is to construct the party wall from two leaves of masonry or timber framing with a cavity between them (see chapter 4: Isolation p. 111). The cavity in the party wall usually extends for the full height of the dwelling, terminating at the underside of the roof. In many houses, the party wall cavity is continuous with the external wall cavity, in order to reduce sound transmission around the edges of the party wall.

Until recently, building regulations in the UK presumed that there was no substantial heat loss through party walls because the dwellings on either side were likely to be heated to a similar temperature. However, research carried out by Leeds Metropolitan University (2007) showed that cavity party walls produced substantial heat loss. The result should not have been unexpected, because it only confirms the normal movement of air in a cavity.

Heat travels by conduction through each leaf of the party wall from the dwelling into the cavity, where it warms the air in the cavity (as illustrated below). The warmed air rises up the cavity, passing out to the atmosphere at the roof junction, and is replaced by cold air drawn in from the connected cavities in the external walls. The overall result is a continuous flow of cold air into the cavity, producing high levels of heat loss.

The solutions are to avoid wall cavities altogether (and rely on other mechanisms to reduce sound transmission) or to fill the cavity with fibrous insulation and seal around its perimeter. Holistic thinking about all parts of the wall's performance is crucial.

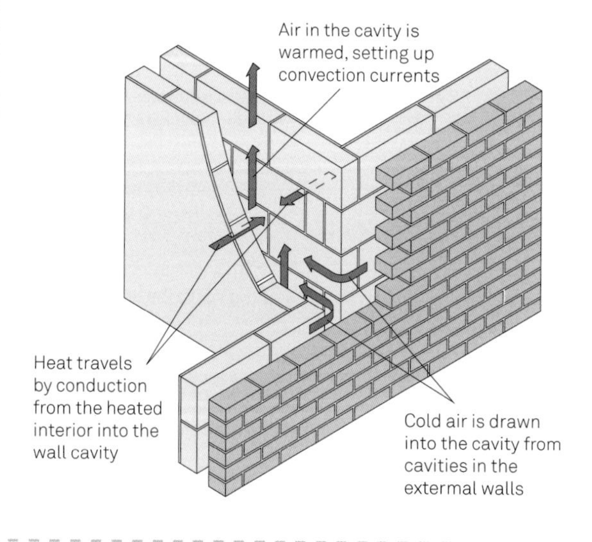

HEAT LOSS IN CAVITY PARTY WALLS

Assessing thermal mass

Having examined how to limit heat transfer through the building fabric, we can now consider how the capacity of the building fabric to absorb energy – its **thermal mass** – affects the building's response to changes of temperature. Thermal mass can be determined using the 'steady state method', which simply calculated the amount of energy needed to raise the temperature of a construction, or by using the 'dynamic method', which assesses the response of the fabric to changing conditions over time.

Steady state method

The κ-value **(kappa-value)** is the amount of energy required to raise the temperature of a square metre of the building element by one kelvin, and is expressed in kJ/m2K (kilojoules per metre squared kelvin). It is the sum of the heat capacities of the layers of which the element is composed, excluding material which lies:

* More than 100 mm from the internal surface (Figure 1–11a) – for example, in an external wall the κ-value would include the thermal mass of the internal finish and some of the masonry. This limit is imposed because only those parts of the element closest to the surface are involved in daily temperature variations.
* More than midway through the element (Figure 1–11b) – for example, only the first 40 mm of an 80 mm thick partition wall would be included in the calculation. This prevents the overestimation of the thermal mass of the internal elements when both surfaces are assessed individually.
* Beyond thermal insulation (Figure 1–11c) – for example, if a wall has internal insulation behind a plasterboard lining, the κ-value would include only the plasterboard.

Table 1.7 shows the κ-values of common constructions. Masonry and concrete constructions have much higher κ-values than timber-framed constructions, except where internal insulation isolates the thermal mass of that masonry.

κ-values are generally used in month-based steady state analysis of energy demand, such as the UK's national calculation methodologies, **SAP** and **SBEM**.

TABLE 1.7 TYPICAL K-VALUES

CONSTRUCTION	K-VALUE (kJ/m²K)
FLOORS	
Suspended timber ground floor, insulation between joists	20
Suspended concrete floor, carpeted	75
Groundbearing concrete slab, screed over insulation	110
WALLS	
Timber-framed wall, one layer of plasterboard	9
Masonry wall, dense plaster, internal insulation, any outside structure	17
Masonry wall, plasterboard on dabs or battens, aircrete block	60
Masonry wall, dense plaster, dense block	190
ROOFS	
Plasterboard, insulated at ceiling level	9

SOURCE: SAP 2012 TABLE 1E. BRE. 2013.

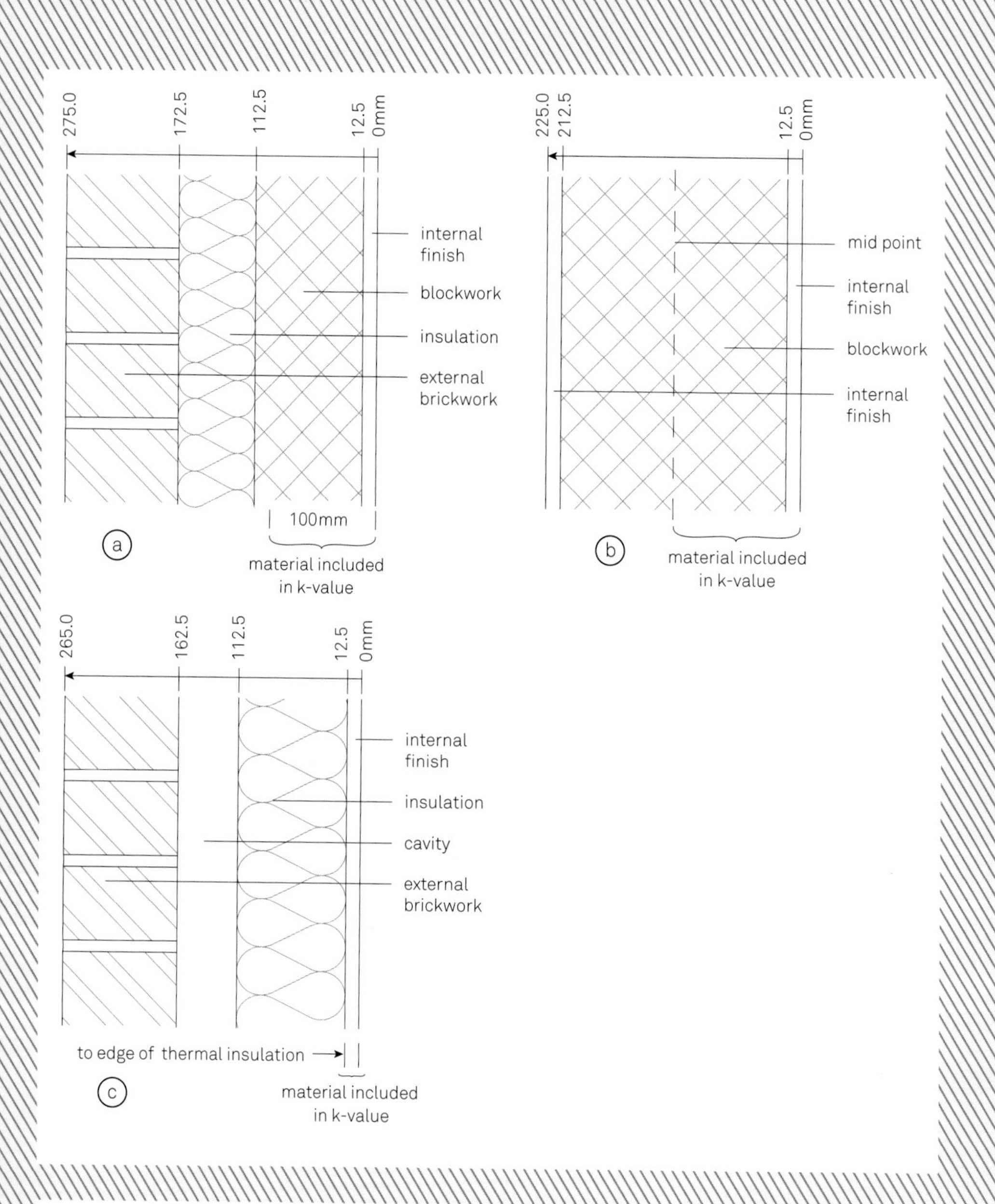

FIGURE 1-11 LIMITS ON MEASUREMENT FOR CALCULATING KAPPA VALUES

Dynamic method

The dynamic approach to thermal mass, commonly referred to as the 'admittance method', uses the density, specific heat capacity and thermal conductivities of materials to calculate measures of thermal mass, which can then be used to assess the thermal performance of elements on an hourly basis.

The dynamic performance of materials can be expressed by:

* **Thermal diffusivity** – a measure of how rapidly heat travels in a material. Under daily temperature cycles, heat will travel further into materials with higher diffusivity, which means they will be more effective for cyclical storage of heat.
* **Thermal effusivity** – the 'thermal inertia' of a material, that is its capacity to absorb and release heat. A material with a high effusivity will be able to store and release a large amount of heat.

Table 1.8 shows that brick and dense concrete have higher diffusivities and effusivities than aircrete, while the thermal insulants have diffusivities in the same range as masonry, but have effusivities which are substantially lower. Brick and concrete will therefore be able to store much more energy than thermal insulants.

TABLE 1.8 DYNAMIC THERMAL PERFORMANCE OF COMMON CONSTRUCTION MATERIALS

MATERIAL	THERMAL CONDUCTIVITY (W/mK)	DENSITY (kg/m^3)	SPECIFIC HEAT CAPACITY (kJ/kgK)	DIFFUSIVITY (m^2/s (x 10^7))	EFFUSIVITY ($Jm^2/Ks^{0.5}$)
Dense concrete block	1.75	2300	1000	7.61	2006
Aircrete block	0.20	700	1000	2.86	374
Brick exposed	0.77	1750	1000	4.40	1161
Brick sheltered	0.56	1750	1000	3.20	990
Dense plaster	0.57	1300	1000	4.38	861
Plasterboard	0.21	700	1000	3.00	383
Wood	0.13	500	1600	1.63	322
Mineral wool batt	0.04	25	1030	14.80	31
PIR board	0.02	30	1400	5.24	30
Steel	50	7800	450	142	13248

SOURCE: BS EN ISO 10456 AND CIBSE GUIDE A3.

The dynamic performance of elements is assessed by analysing a notional 24-hour heating and cooling cycle to determine the key properties of admittance, decrement and **surface factor**.

Admittance

Admittance, Y, (W/m²K, watts per metre squared kelvin) is a measure of how rapidly heat will pass between the surface of an element and the interior of the building. Admittance is most strongly affected by the part of the element nearest the surface. Figure 1–12 shows the U-values and admittances for different thicknesses of three typical materials: one dense (concrete); one medium weight (blockwork); and one lightweight (insulation). The calculations are based on the values in table 1.9. Although the dense and medium weight materials have much higher admittances than the lightweight material, in each case the admittance is practically constant beyond a certain thickness. This stability justifies the use of the 100-mm thickness limit in the steady state κ-value calculations.

TABLE 1.9 PROPERTIES OF MATERIALS IN FIGURE 1-12

	DENSITY (kg/m³)	CONDUCTIVITY (W/mK)	SPECIFIC HEAT CAPACITY (J/kg m³)
Concrete	2500	1.5	1000
Blockwork	1000	0.25	1000
Insulation	25	0.03	1000

SOURCE: BASED ON DATA FROM BS EN ISO 10456 AND CIBSE GUIDE A3.

Table 1.10 shows admittance values for typical constructions. Those with higher admittances will absorb more heat from the occupied space than those with lower admittances, and so will be better for reducing cooling loads (see: Solar gain p. 31).

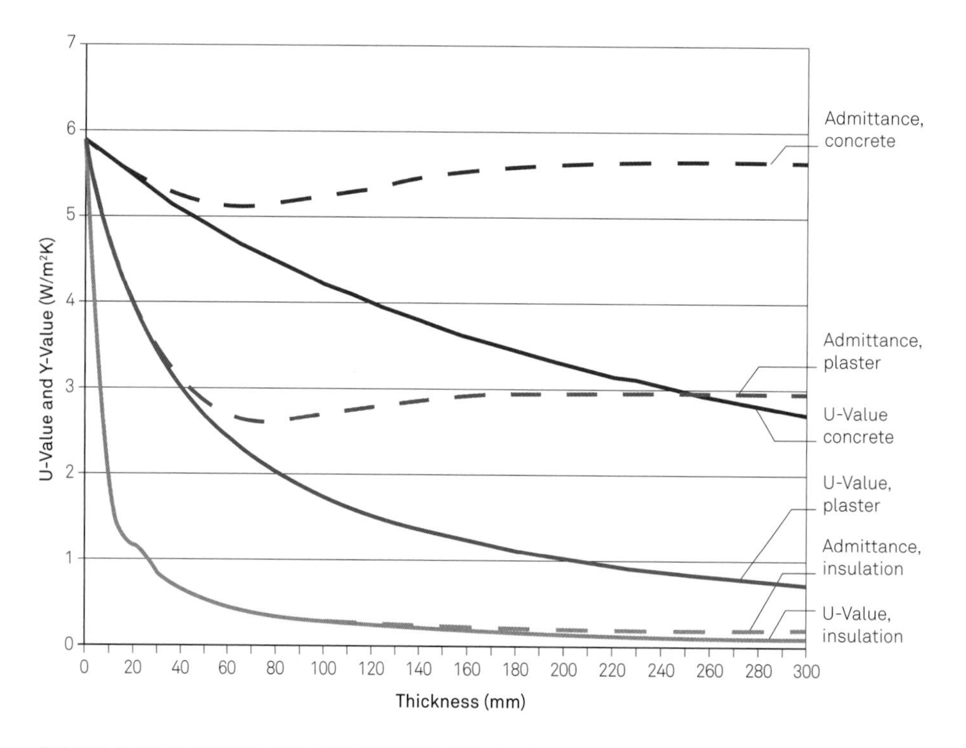

FIGURE 1-12 U-VALUES AND ADMITTANCES FOR THREE COMMON MATERIALS

TABLE 1.10 TYPICAL DYNAMIC THERMAL MASS PROPERTIES FOR WALLS					
CONSTRUCTION	U-VALUE (W/m^2K)	k-VALUE (kJ/m^2K)	ADMITTANCE (W/m^2K)	DECREMENT FACTOR	DECREMENT DELAY (HR)
103 mm brick 25 mm cavity 75 mm PIR 100 mm aircrete 12.5 mm plasterboard on dabs	0.21	55	1.86	0.24	11.42
103 mm brick 25 mm cavity 75 mm PIR 100 mm dense block 13 mm gypsum plaster	0.25	214	5.55	0.21	10.36
103 mm brick 50 mm cavity 19 mm plywood sheathing 140 mm mineral fibre between studs 12.5 mm plasterboard	0.23	10	0.79	0.54	6.61

SOURCE: BASED ON DATA FROM BS EN ISO 10456 AND CIBSE GUIDE A3.

Decrement and decrement delay

The outside surface temperature of an element will vary over the course of a day as a result of changing air temperature and radiative losses and gains. Some of the heat at the surface will travel through the element and produce temperature variations at the internal surface. However, the temperature range at the internal surface will be smaller than that at the external surface, because the thermal mass of the structure acts as a damper on the heat flow. This is illustrated in Figure 1–13.

The effect of thermal mass on the temperature change is measured by the **decrement factor, f**, which is the ratio of the internal temperature range (minimum to maximum) to the external temperature range. For example, a wall which experiences a 20°C temperature swing on the outer face, but only a resulting 5°C swing on the inner face, would have a decrement factor of 0.25. The time taken for heat to pass from one side of the structure to the other is the **decrement delay**, which is measured in hours. Examples of decrement factor and decrement delay for typical constructions are shown in table 1.10 above.

The decrement factor and decrement delay are used to assess the impact of external temperature on the building. Where the diurnal temperature range is substantial, constructions with low decrement factors are desirable: the thermal mass of the structure will moderate temperature extremes, minimising internal gains and losses.

A decrement delay of nine or ten hours will ensure that the effect of high external temperatures (which occur at the middle of the day) will not reach the building interior until the cooler night, at a time when the rise can be mitigated by ventilation cooling (see chapter 2: Ventilation p. 60).

Surface factor and surface delay

The **surface factor, F,** compares the cyclical variation in heat flow from a surface to the cyclical variation into the surface. (It is ridiculous that the decrement factor is f and the surface factor is F. Sometimes building physics seems designed to be confusing!) The surface factor is used to assess the effect of solar radiation and internal gains on internal surfaces. Structures with high admittances will have low surface factors. The time lag between the peak flows is defined by the **time factor,** ψ (psi).

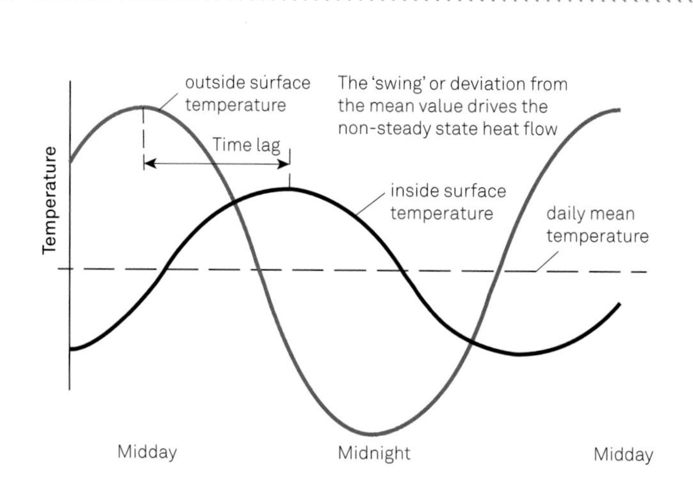

FIGURE 1-13 DAILY VARIATION IN TEMPERATURE THROUGH A WALL

Mass-enhanced U-values

U-values are usually calculated independently of thermal mass, but if it is likely that the exterior temperature will fluctuate above and below the internal temperature of a building over the course of a day, a more accurate picture of performance can be gained by taking thermal mass into consideration (rather than using the basic steady state U-value calculation).

As the balance of internal and external temperature changes, so the direction of heat flow through the element will change. At night, when the external temperature is lower, heat transfer will be from inside to outside, but the transfer will be slowed down by the thermal mass of the structure. During the day, if the internal temperature is lower than outside, the direction of heat transfer will reverse, and some of the heat stored within the construction will travel back to the building interior. Taken together, those effects will give a lower effective U-value.

The improvement in performance produced by a mass-enhanced U-valve is directly related to the decrement delay: where there is a wide diurnal temperature variation, a structure with higher thermal mass will transfer less heat and so be more energy efficient.

Whole building energy performance

Having considered the main interactions of heat with the building fabric we can now examine how they can be applied in order to design buildings which provide suitable internal conditions with minimum energy use. We will begin by considering how to control solar gain, then examine how to minimise heating demand, and finally look at how to avoid overheating while minimising cooling demand.

Solar gain

In temperate climates solar gain is generally beneficial during the heating season because it can supply part of the heating demand. However, it can be problematic in other seasons – potentially raising internal temperatures to uncomfortable levels. In other climate zones solar gain may be problematic in all seasons. But irrespective of whether solar gain is beneficial or problematic, there are three fundamental considerations:

* The amount of solar radiation at the site, which is determined by the building's location
* The amount of that radiation reaching the openings, which is determined by the orientation of the building and the shading of the openings
* The amount of solar radiation which passes through the openings to the interior, which is determined by the characteristics of the glazing (see above: Controlling heat transfer through openings p. 20)

In some projects it will also be necessary to control indirect solar gain through the opaque fabric.

Location

The primary determinant of the amount of solar radiation a building can receive during a year is its latitude (see chapter 5: The sun).

The amount of solar radiation received on the facets of a building will vary with their orientation. Figure 1–14 shows the relative amount of solar energy falling on vertical surfaces at different compass directions for a site at 55°N (e.g. northern England/southern Scotland). The heating season values in this example show a large bias towards SE-S-SW compared with NW-N-NE, when the sun is low on the horizon. The differences are not as great during the summer when the sun has the greatest arc. Solar radiation will also be more intense during the summer, because the sun is higher in the sky.

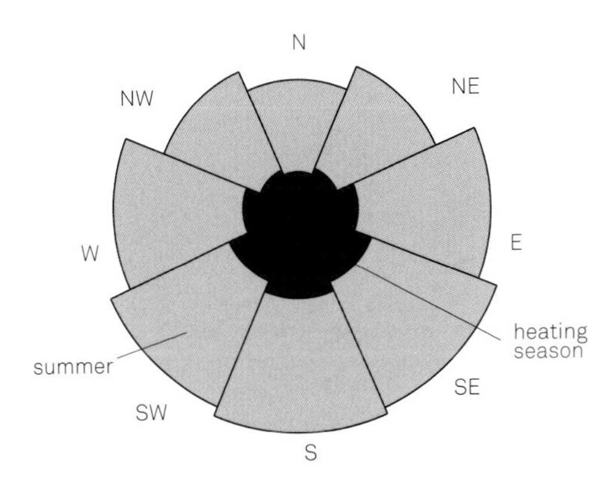

FIGURE 1-14
RELATIVE PROPORTION OF SOLAR ENERGY BY COMPASS DIRECTION
Data derived from SAP 2012

Seasonal differences are more marked at more northerly/southerly latitudes: at the equator the sun is directly overhead at noon, with little variation in its path or day length through the year. The amount of available solar radiation will also be affected by the climate, as cloud cover will reduce the amount of radiation reaching the ground.

Orientation and shading

The proportion of the available solar radiation a building receives will be affected by the orientation of openings and their shading. For a building with the solar potential shown in Figure 1–14 the maximum benefit of solar gain in winter would be obtained by setting a high proportion of the area of openings to face between south-east and south-west. However, adopting that orientation pattern might not be beneficial over the whole year, because the reduction in heating energy demand produced by solar gain in the winter might be outweighed by cooling energy demand during the summer.

The change in the angle at which solar radiation reaches the earth (see chapter 5, Figure 5–03 p. 122) can be used to maximise the benefit of solar gain in winter, while minimising gain in the summer. For example, as Figure 1–15 shows, in the winter (when the sun is at its lowest), solar radiation will penetrate deep into a building, while in the summer, (when the sun is higher), direct solar radiation can be blocked by shading the opening with an overhang (such as a deep soffit on a pitched roof) or brise soleil. External blinds will also reduce solar gain, but internal blinds and curtains will be less effective, because the solar radiation will already have reached the interior of the building.

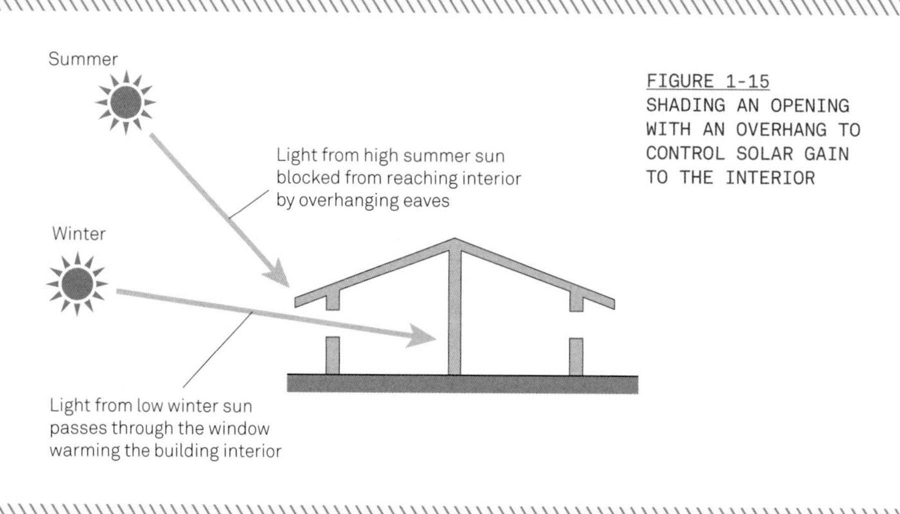

FIGURE 1-15
SHADING AN OPENING WITH AN OVERHANG TO CONTROL SOLAR GAIN TO THE INTERIOR

Reducing heating demand

Figure 1–16 is a simplified schematic diagram showing the relationship between the main factors which go towards ensuring the building is maintained at the design temperature. (One simplification is the omission of energy for artificial lighting, which is affected by the sizing of the windows, but is, in turn, linked to the rate of heat loss through the openings and the amount of solar gain.) Reducing the heat losses on the left-hand side of the diagram, by minimising heat transfer through the fabric and through air movement, will reduce heating demand. However, on the right-hand side of the diagram, maximising the beneficial effect of solar gain will reduce the heat demand.

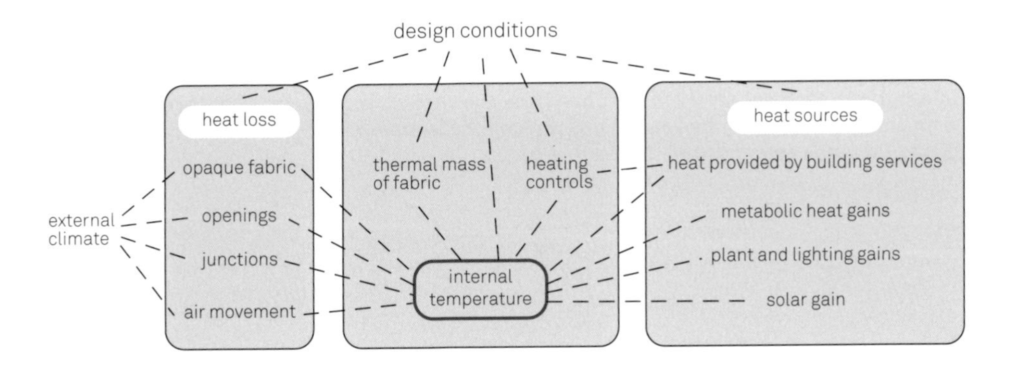

FIGURE 1-16 FACTORS AFFECTING HEAT DEMAND

The building needs adequate controls to ensure that the heating system responds promptly to the changing temperature inside the building. The complexity of the heating system and its controls will, necessarily, vary with size and complexity of the building and the size of the acceptable temperature range. (However, the design and specification of heating systems and their controls is beyond the scope of this book.)

Finally, to optimise energy demand it is crucial to consider the role of thermal mass.

Thermal mass and heating

When a building is heated, a part of the energy will raise the temperature of the fabric, rather than the building's interior air spaces, and fabric with high thermal mass will absorb more energy than fabric with low thermal mass. A building with high thermal mass will therefore take longer to reach the desired temperature. Similarly, when the heating is turned off the building with high thermal mass will cool more slowly, as heat will be released from the fabric into the building interior.

Specifying fabric with high thermal mass helps to maintain a steady internal temperature because it dampens slight variations: if the temperature starts to drop, the structure will release some of the stored energy back into the building's interior. Conversely, if the temperature starts to rise further the structure will absorb some of the energy. (A useful analogy for this aspect of thermal mass is the momentum of vehicles: a heavy lorry will require more energy and a longer time to reach 80 km/h than a car, but is easier to keep at a steady speed.) Fabric with high thermal mass is therefore suited to buildings which are continuously occupied and maintained at a similar temperature throughout the heating season (e.g. hospitals).

The effect of thermal mass on heating demand in intermittently heated buildings depends on the heating pattern and the fabric heat loss. In moderately well-insulated dwellings (which could be characterised by a wall U-value of 0.25 W/m^{2}K) which are heated twice a day, low thermal mass minimises the amount of energy required to bring the building up to temperature in the morning and evening, although it also means the buildings will cool rapidly once the heating is turned off. This represents a typical domestic heating pattern for the UK: a few hours around breakfast and a longer period in the evening.

However, if the dwelling is very well insulated (characterised by a wall U-value of 0.15 W/m^{2}K) the amount of heat lost (and therefore the temperature drop) between heating periods will be much less, and high thermal mass will be help to minimise the temperature drop and require less energy over the daily heating cycle.

Reducing cooling demand

Figure 1–17 is a simplified schematic diagram showing the relationship between the main factors associated with overheating. On the left-hand side the climatic conditions result in heat transfer into the interior through air movement and through the fabric, while solar radiation results in heat gains through openings and through the fabric.

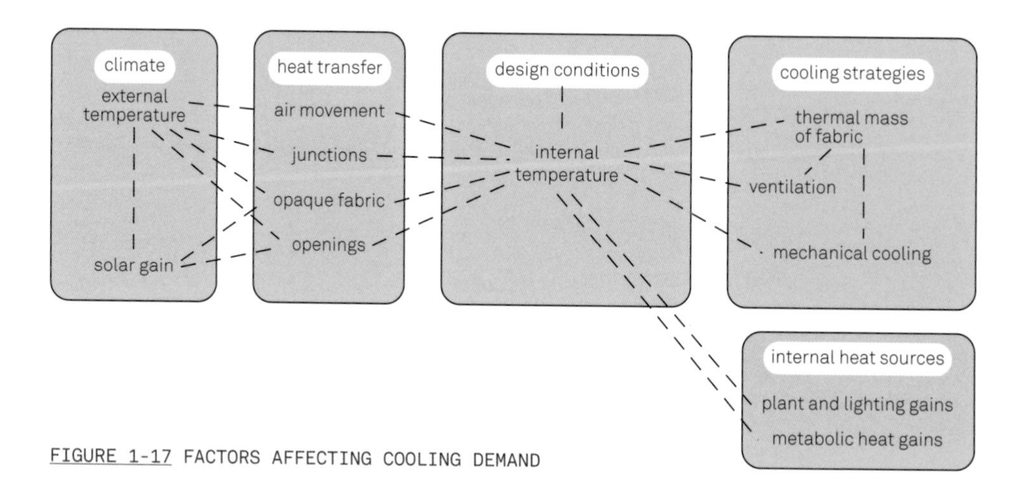

FIGURE 1-17 FACTORS AFFECTING COOLING DEMAND

The risk of overheating can be mitigated, and the need for mechanical cooling reduced or even eliminated by:

* Minimising solar gain through the opaque fabric and openings
* Reducing heat transfer through the fabric from outside to inside[6]
* Using ventilation systems to remove warmed air from the building and introduce cooler air (see chapter 2: Air for moderating temperature)
* Making use of the thermal mass of the fabric

The rest of this section considers the measures available to minimise solar gain, and how thermal mass may be employed to reducing overheating.

Overheating through opaque fabric

A proportion of the solar radiation falling on the opaque external surfaces of a building will be absorbed by that surface, raising its temperature (see above: Heat and light p. 9). Over time, the heat generated at the surface will travel into the building, eventually reaching its interior and warming it by radiation and convection. The potential for overheating can be addressed, firstly by reducing the amount of radiation absorbed at the surface and secondly by slowing the rate of heat transfer from the external surface to the building interior.

The proportion of incident solar radiation which is absorbed is determined by the reflectance of the surface, because surfaces with high reflectance will also have low absorptivity, given that, when there is no transmission of radiation, the total of reflectance and absorptivity will be 1. For example, surfaces with high reflectance, such as white-painted surfaces, will absorb less solar

radiation than those with low reflectance, resulting in less heat gain. 'Cool roofs' in the USA use highly reflective solar coatings to minimise solar gain (This is hardly a new strategy, as the white-washed villages of Greece demonstrate.)

Although solar coatings are typically white or metallic, coloured 'cool' coatings are available which have higher reflectances than standard coatings of the same colour. In most cases it will be beneficial for the surface to have a high emissivity, so that it re-radiates much of the heat it receives.

The specification of a thermally massive structure with a high decrement delay (see above Decrement and decrement delay), say, upwards of nine or ten hours, will slow the rate at which heat is transferred to the building interior. The heat will reach the interior at night, when the air is cooler, and the heat may be removed by night-time ventilation, readying the building for the next day. In contrast, a thermally light structure, with a low decrement delay (say, three or four hours) will quickly transmit heat to the building interior. The internal surface temperature will rise at the same time as the air temperatures are highest, increasing the risk of overheating. Indeed, it is difficult to think of a situation where a thermally light structure would be beneficial in addressing indirect solar gain. A solar collector (see chapter 2: Beneficial air movement in air spaces p. 52) would also reduce the amount of solar gain at the surface that could be transferred through the fabric.

Overheating through openings

Solar radiation transmitted through the glazing of openings will fall on a surface inside the building, warming the fabric and thus the space. This solar gain will be beneficial during the heating season, because it will reduce demand on heating plant, but during the summer can result in excessive internal temperatures. The amount of solar radiation reaching the building interior can be reduced by addressing the orientation of openings, shading and glazing, but in many cases solar gain will still result in overheating. Measures to reduce solar gain may also adversely affect daylighting.

The risk of overheating can be mitigated by specifying constructions with high thermal mass which will absorb the solar energy during the day and release it at night when the air temperature drops. Constructions of dense masonry and concrete will provide the best absorption. The soffits to internal floors should be exposed (suspended ceilings will tend to isolate the thermal mass from the internal surface), while profiled soffits will increase the surface area and improve the absorption of heat into the fabric.

For the thermal mass to be effective in countering the overheating problem, the fabric must be cooled at night to release the heat absorbed during the day (see chapter 2: air for moderating temperature p. 57).

Phase change materials

When a solid material is melted, energy is required to raise the material to its melting point, but a much greater amount of energy is used to make the phase change from solid to liquid. That energy, which is known as **latent heat**, may be provided deliberately or be drawn from the materials surroundings: for example, an ice cube dropped in a drink will melt as it gains energy from the liquid by conduction; it will also cool the drink.

The same principle can be applied to reduce the overheating of buildings through the use of **phase change materials** (PCMs). These are typically paraffin waxes or salt compounds which are engineered to melt at a specific temperature. They are supplied in sheet format or incorporated into gypsum board or a similar finishing material.

A PCM behaves as a normal material as long as the internal temperature does not reach its melting point. If the internal temperature exceeds the melting point, the PCM will absorb heat from the space as it melts, so limiting further temperature rise in the space. In effect, the PCM provides additional thermal mass at temperatures above its melting point. Correctly configured a PCM can provide the same thermal mass as five times its thickness of concrete.

Figure 1–18 shows the temperature profile of a room with and without a PCM installed. The temperature with the PCM never reaches the same peak as that without the PCM, but also it never reaches the same low point. This is because once the room temperature drops below the melting point of the PCM it will start to solidify and will release heat back into the room.

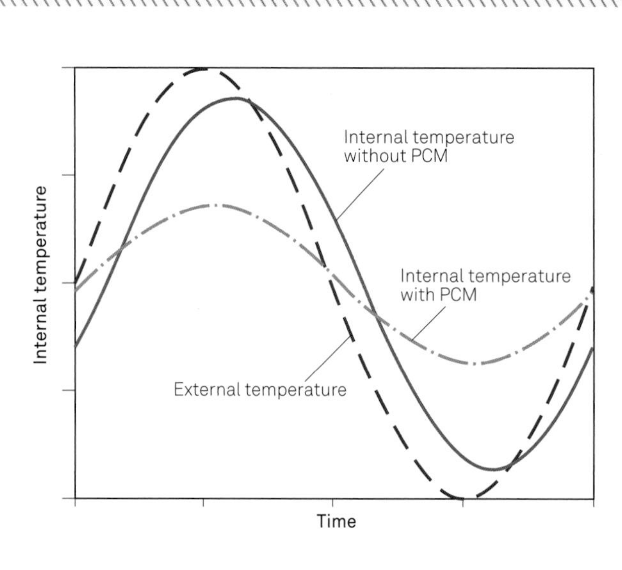

FIGURE 1-18
THE EFFECT OF A PHASE CHANGE MATERIAL ON INTERNAL TEMPERATURE

For a PCM to be used successfully:

* The melting point must be matched to the predicted internal conditions: too low and it will increase the heating load; too high and it will not be activated
* There must be adequate night-time cooling to solidify and 'recharge' the PCM for the following day

PCMs are only beneficial for buildings with low thermal mass: by reducing overheating they reduce or even eliminate the cooling load.

Modelling whole building energy performance

Analysis of building energy performance during the design process should be carried out using software which considers the internal and external conditions of the building, its size and layout, the thermal performance of the fabric and the performance of the building services (including the efficiency, responsiveness of the controls), including any renewables. There are two main approaches to analysis:

* Steady state methods, which typically model the building's performance on a monthly basis,[7] by calculating the energy required to maintain the design conditions for that period. They use steady state values for thermal mass (see chapter 3: Steady state analysis).
* **Dynamic simulation modelling** (DSM) which models the building's state at hourly intervals, with one hour's results forming the starting point for the next. DSM uses dynamic thermal mass values (see: Dynamic method in Assessing thermal mass) and is more accurate than steady state methods. DSM requires more data on internal and external environmental conditions.

The predicted energy performance derived from modelling will never exactly match the actual performance of the building when in use: in practice the actual performance is invariably worse than the predicted performance. Part of that 'performance gap' will be the inevitable result of simplifications built into the model and assumptions made about the occupants' behaviour.

However, larger discrepancies can result from:

* Inaccurate data input during analysis
* Poor site practices during construction
* Incorrectly set controls during use
* Poor communication with building users about the operation of the building

Addressing those issues is, inevitably, more difficult than identifying them.

Heat in the bigger picture

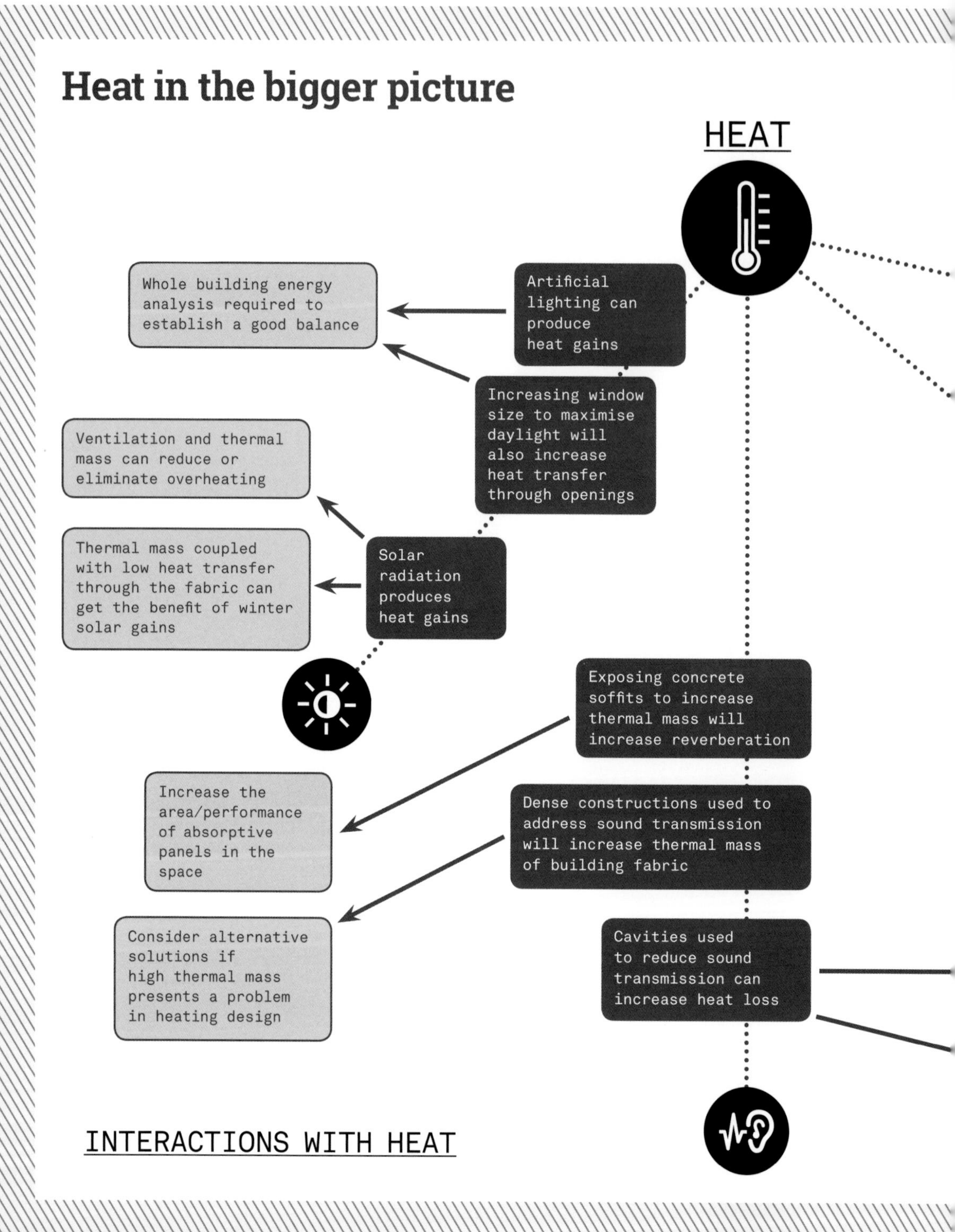

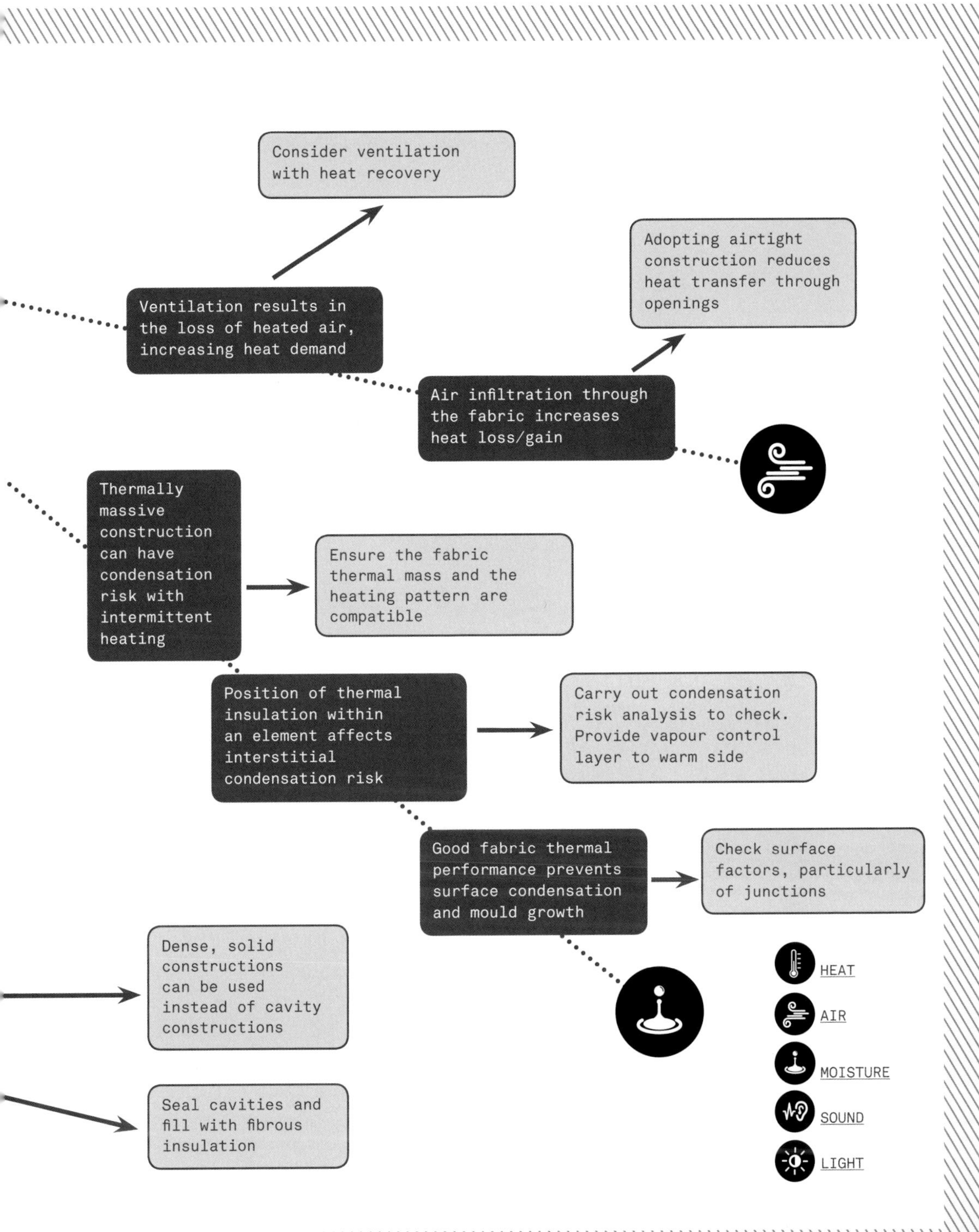
Consider ventilation with heat recovery
Adopting airtight construction reduces heat transfer through openings
Ventilation results in the loss of heated air, increasing heat demand
Air infiltration through the fabric increases heat loss/gain
Thermally massive construction can have condensation risk with intermittent heating
Ensure the fabric thermal mass and the heating pattern are compatible
Position of thermal insulation within an element affects interstitial condensation risk
Carry out condensation risk analysis to check. Provide vapour control layer to warm side
Good fabric thermal performance prevents surface condensation and mould growth
Check surface factors, particularly of junctions
Dense, solid constructions can be used instead of cavity constructions
Seal cavities and fill with fibrous insulation
HEAT
AIR
MOISTURE
SOUND
LIGHT

02

Air

DEFINITION:
AIR: THE MIXTURE OF MATTER IN GASEOUS AND VAPOUR STATE OF WHICH THE EARTH'S ATMOSPHERE IS COMPOSED

Air is fundamental for human survival: without air or, more accurately, without oxygen, we suffocate and die. We can also be poisoned by other gases, and by small particles present in air (if the concentrations are high enough). Outdoors (and away from sources of pollution) where the air consists of the standard mixture of gases there is no question of having the right balance of gases. But inside buildings, human beings and their activities result in higher levels of harmful gases, water vapour and pollutants – including the products of combustion, toner gases from photocopiers and even unpleasant odours.

The need to ensure a suitable air supply is not restricted to building occupants: many space and water heating appliances require oxygen for combustion. There are also conditions in which the building fabric needs to be properly ventilated.

Until recently, much of the air supply for the building occupants and for combustion appliances was provided by what was euphemistically referred to as ‘fortuitous ventilation’: that is, the infiltration of air through gaps and cracks in the fabric. Such air movement is usually inefficient (resulting in heat loss as warmed air is replaced by cold external air) and often uncomfortable (producing uneven internal temperatures and draughts).

We need to design buildings which have an adequate air supply and are also energy efficient and comfortable. That requires an understanding of the behaviour of gases and the broader influences of the earth's atmosphere and the forces which can affect the structure of buildings, but can also be used to drive the ventilation of buildings.

This chapter considers the requirements of an air supply for the wellbeing of building occupants, as well as the effects of air supply on the building fabric. It also considers the various ventilation strategies available to designers. Throughout the chapter we will keep in mind the interactions between air – particularly in ventilation – and heat and moisture.

The fundamentals of air

In order to understand the behaviour of air in and around buildings we need to consider how gases behave as their temperatures and pressures change, and how that in turn affects the movement of air in the earth's atmosphere. Then we need to examine how that movement, in the form of wind, produces forces which act on the surfaces and structures of buildings. Finally, we will look at how air movement within a building is produced by differences in air pressure and temperature, including the **stack effect**.

The behaviour of gases

Gas is one of the four states of matter: the atoms and molecules in gases move freely and are separated by much greater distances than those of solids and liquids. Despite that separation there are constant collisions between gas particles and between the particles and adjacent surfaces. The force of those collisions exerts a **pressure** within the gas and on those surfaces. Pressure is measured in pascals (Pa), but when dealing with atmospheric pressure we usually work in kilopascals (kPa), which are a thousand times larger. Atmospheric pressure is also measured in bar, equivalent to 100 kPa, and millibar, equivalent to 100 Pa.

The pressure exerted by a quantity of gas is affected by its volume and temperature (see Figure 2–01): heating a fixed amount of any gas increases the internal energy of the molecules, which will either result in an increase in volume (expansion) as the molecules move further away from each other or, if expansion is not possible (because the gas is contained) an increase in pressure. Conversely, cooling the gas will reduce either the volume or the pressure. As the pressure and temperature of a gas are connected, it is often necessary to look at the behaviour at a **standard temperature and pressure** (STP). Usually this is 0°C and 100 kPa.

The relationships between the volume, temperature and pressure of a gas are predicted by the **gas laws**. In conditions close to STP most gases conform closely to the gas laws and their behaviour is easy to predict, but at higher temperatures and lower pressures their behaviour diverges from the laws.

Where there is a mix of gases – as there is in the earth's atmosphere – each gas in the mixture exerts part of the pressure; this is known as its **partial pressure**. The total pressure exerted by the gas mixture is the sum of the partial pressures of all the gases in it. Where there is water vapour present in the gas mixture, the total pressure is the sum of the gas pressure and the water vapour pressure.

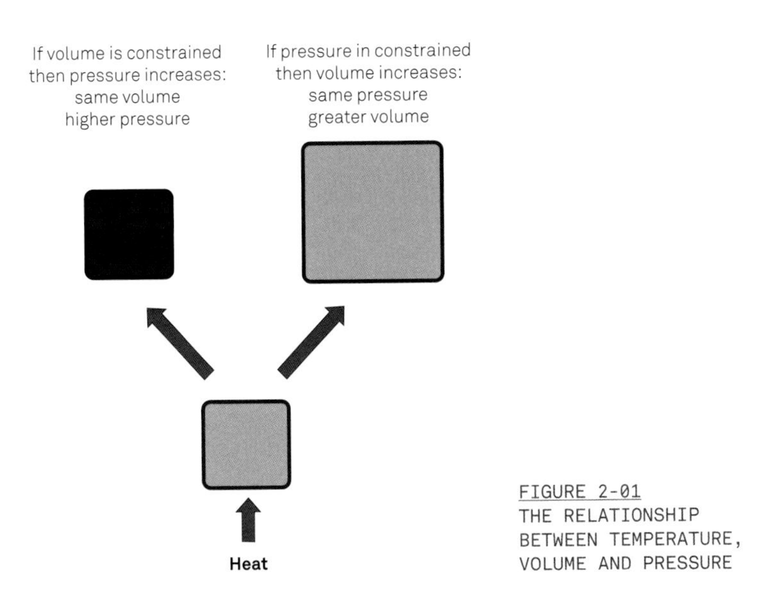

FIGURE 2-01
THE RELATIONSHIP BETWEEN TEMPERATURE, VOLUME AND PRESSURE

Variations in air temperature and pressure result in air moving from areas of high pressure to low pressure: that occurs on a large scale in the earth's atmosphere, and on a smaller scale in and around buildings.

The earth's atmosphere and the weather

The lowest layer of the earth's atmosphere – the **troposphere** – extends approximately 12 km from the surface of the earth and exerts a pressure on the earth which is roughly equivalent to the weight of the air above it. Atmospheric pressure is greatest at sea level and decreases with altitude, and is affected by temperature and humidity. Differences in atmospheric pressure result in winds, as air moves from higher to lower pressure areas.

Large-scale wind patterns are affected by topography, as mountains and valleys produce variations in wind direction and speed. At a smaller scale, sea breezes at coasts change direction according to the relative temperatures of sea and land during diurnal cycles of warming and cooling. The operation of all these forces produces the climatic conditions that act upon a building, including wind, rainfall and, to some extent, temperature.

Wind action

When wind blows onto a building the air is mainly deflected around it, producing different air pressures across the building, with high pressures on the windward face and lower pressures on the other faces (Figure 2-02). The pressure differentials result in forces acting on the fabric of the building, usually referred to as **wind loads**, and also in pressure differences between the inside and outside which can drive natural ventilation (see: Natural ventilation p. 61).

The pressure differences, and therefore the wind loads, are determined by the speed of the wind and its direction, and the orientation and configuration of the building relative to the wind. The presence of the building will also affect the wind forces experienced around the building, in some cases creating strong air currents that affect pedestrians.

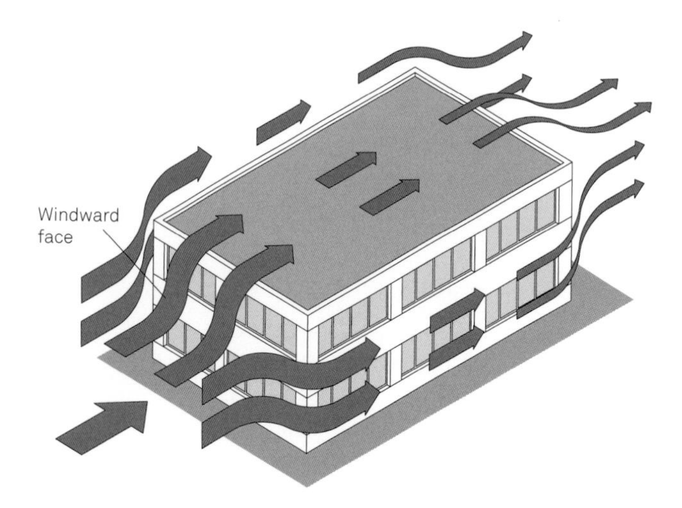

FIGURE 2-02
AIR CURRENTS PRODUCED BY THE WIND BLOWING ACROSS A BUILDING

Wind speed

As the speed of the wind at a site constantly varies, wind speeds are quoted as standardised speeds for a height of 10 m above ground blowing across open country. That standardised wind speed is then adjusted to site conditions, taking account of:

* The terrain – effective wind speeds will be lower in urban and suburban areas than rural areas
* Altitude – wind speeds are generally greater at higher altitudes
* Position relative to orographic[1] features such as hill and escarpments, and proximity to the sea

Wind speed can be expressed in several different ways:

* Ten-minute mean wind speeds with a 50-year return period (that is, there is a 2% probability of that speed being exceeded in any one year) are used in structural calculations because they express the greatest forces likely to be encountered
* Mean annual wind speeds are representative of conditions experienced throughout the year and are therefore used in ventilation

For any location, the wind speeds used for structural calculations will be many times greater than those used in ventilation calculations.

Wind direction

The wind speed at a site will vary with the wind direction. The analysis of wind loads and ventilation pressure differences should therefore consider the forces produced when the wind blows from each the eight cardinal wind directions (N, NE, E, SE, S, SW, W, NW). For larger or more sensitive buildings (e.g. a naturally-ventilated commercial building) it may be necessary to calculate forces for every 30° segment.

Most sites have a prevailing wind direction (the direction the wind blows most frequently), which is affected by large-scale weather patterns and local topography. The wind forces produced by the prevailing wind are of particular importance because they represent the commonest conditions the building will experience.

Wind loads and forces

Once the site wind speed has been calculated for eight or twelve directions the air pressures and wind loads acting on the building can be calculated. The methods vary for different types of structure (and for the assessment of ventilation forces), but in outline:

1. The wind speed is used to calculate a peak velocity pressure (or the 'dynamic pressure', which is effectively the same thing), taking account of turbulence produced by neighbouring buildings, terrain and exposure.
2. The peak velocity pressure is multiplied by an external pressure coefficient (C_{pe}) or an internal pressure coefficient (C_{pi}) to obtain the pressure on the external or internal surface.
3. Finally, the pressure on a surface is multiplied by its area to obtain the wind load on that surface.

The wind forces acting on the building depend on the size and shape of the building and its position relative to the wind direction. Windward surfaces, such as walls facing directly into the wind, will experience positive pressures and therefore positive wind loads, while the sides and leeward faces will experience negative pressures and therefore negative wind loads (that is, suction). This is illustrated in Figure 2–03, which shows the varying intensity of air pressures on the same simple building illustrated in Figure 2–02.

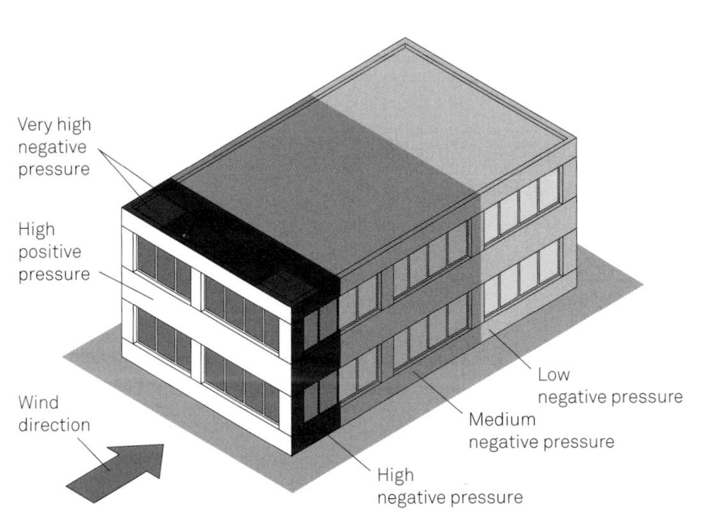

FIGURE 2-03
AREAS OF POSITIVE AND NEGATIVE PRESSURE PRODUCED BY WIND ACTION ON A BUILDING'S SURFACES

The stack effect

The stack effect occurs when a column of air – such as the air in a chimney – is heated. The warmed air expands, becoming less dense, and is displaced upward by colder denser air (the same mechanism of convection; see chapter 1: Mass transport p. 5). The warmed air rises up the column, while colder air is drawn in at the base of the column, producing a current of rising air (Figure 2–04).

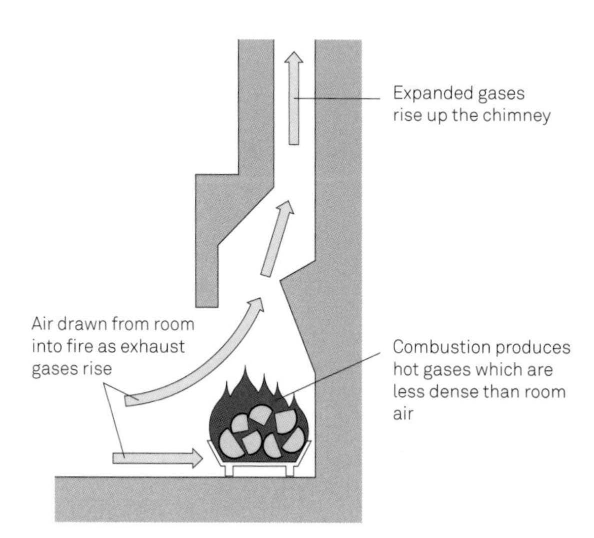

FIGURE 2-04
THE STACK EFFECT IN A CHIMNEY

The stack effect occurs in chimneys and flues, where combustion provides the heat, and in any part of a building in which there is a column of air – for example, a stairwell or atrium – which may be heated by solar gain or gains from the building occupants and their activities, as illustrated in Figure 2–13, p. 64.

The volume of air moved by the stack effect depends on:

* The stack height – taller stacks generate more air movement because there is a greater pressure difference between the top and bottom of the stack[2]
* The temperature difference between the stack (average stack temperature) and outside air – a greater temperature difference results in more air movement, so more air will move during the heating season when the temperature difference is greater, than during the cooling season
* The free area of openings – larger openings at the top and bottom of the stack will result in a greater volume of air being moved

Air movement in and out of openings on the same side of the building can be produced when internal and external temperatures are different. As Figure 2–05 shows, where the internal temperature is higher, the internal air pressure will be greater than the external air pressure at the top of a space, while the external pressure will be greater at the bottom of the space. As a result, air will move out of the building at the upper opening and into the building at the lower opening. Where the external temperature is greater the air flows are reversed. This phenomenon is one of the drivers of single-sided ventilation (see: Single-sided ventilation p. 46).

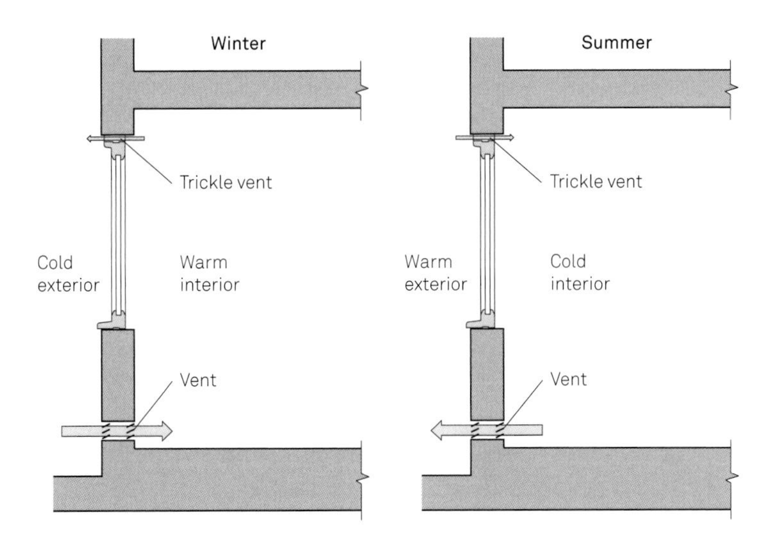

FIGURE 2-05
PRESSURE DIFFERENCES DRIVING SINGLE-SIDED VENTILATION

CONDITIONING THE COMMONS

One of the earliest air conditioning systems – which supplied the temporary House of Commons between 1836 and 1852 – was driven not by mechanical fans but by the stack effect. The system, designed by David Reid, was powered by a large furnace and a tall chimney located adjacent to the chamber of the Commons. The updraught generated by the chimney was used to extract air from the chamber, resulting in fresh air being drawn in through a multitude of small holes in the floor. The chamber could be warmed by passing the incoming air over hot water coils, or cooled by suspending blocks of ice in the air inlets.

The conditions in the temporary House of Commons were vastly superior to those in the previous building, leading one MP to comment: 'To the skill, zeal and determination of Dr Reid, it is owing that the members of the House of Commons can now pursue their senatorial duties without a sacrifice of health or comfort.'

Air and the building fabric

Wind forces can damage the building fabric;[3] they can also result in the infiltration of air through the building fabric with detrimental effects to the comfort of occupants and the energy efficiency of the building. However, there are also conditions where it is necessary to induce air movement in air spaces within building elements to remove high concentrations of moisture or to take advantage of warmed air in the fabric.

The structural effects of wind action

Wind loads on buildings can result in components of the building being dislodged or damaged. Problems most commonly occur on roofs, which are subject to strong suction forces, particularly at perimeters. Fixing specifications for roof coverings should take account of the varying wind loads across the roof (see Figure 2–03 p. 45): the restraints for sheet or membrane coverings must also resist the wind forces acting on layers such as insulation beneath the weatherproofing layer.

Extreme wind forces can produce failure of the building structure, resulting in the collapse of all or part of the building. Although that is uncommon in the UK, there are many parts of the world where extreme weather events, such as hurricanes and tornadoes, do destroy buildings. Although the occurrence of such events is not predictable, it is not wholly random, because the conditions which give rise to them are produced by the large-scale global weather systems (see above: The earth's atmosphere and the weather p. 43). Tornado Alley gets its name for a reason.

In order to protect occupants, buildings should be designed and constructed to withstand local environmental hazards.

Infiltration

Infiltration is the unintended movement of air between the exterior and interior of a building through the fabric. (Ventilation, by contrast, is the deliberate provision of fresh air, which may be conditioned and the removal of exhaust air.) Infiltration occurs at gaps and cracks in the building fabric, particularly at junctions between building elements and around openings (see Figure 2–06). It is driven by wind action and the stack effect, which can move large amounts of air into and out of buildings, causing draughts which affect user comfort and significant heat loss through mass transfer (see: chapter 1 Mass transfer p. 5).

Until relatively recently, infiltration was included in the overall air supply for a building when assessing the ventilation requirements, However, it is not possible to meet current and future standards of energy efficiency without minimising infiltration and providing the necessary ventilation in a deliberate controlled way. The mantra 'build tight, ventilate right' sums up the strategy.

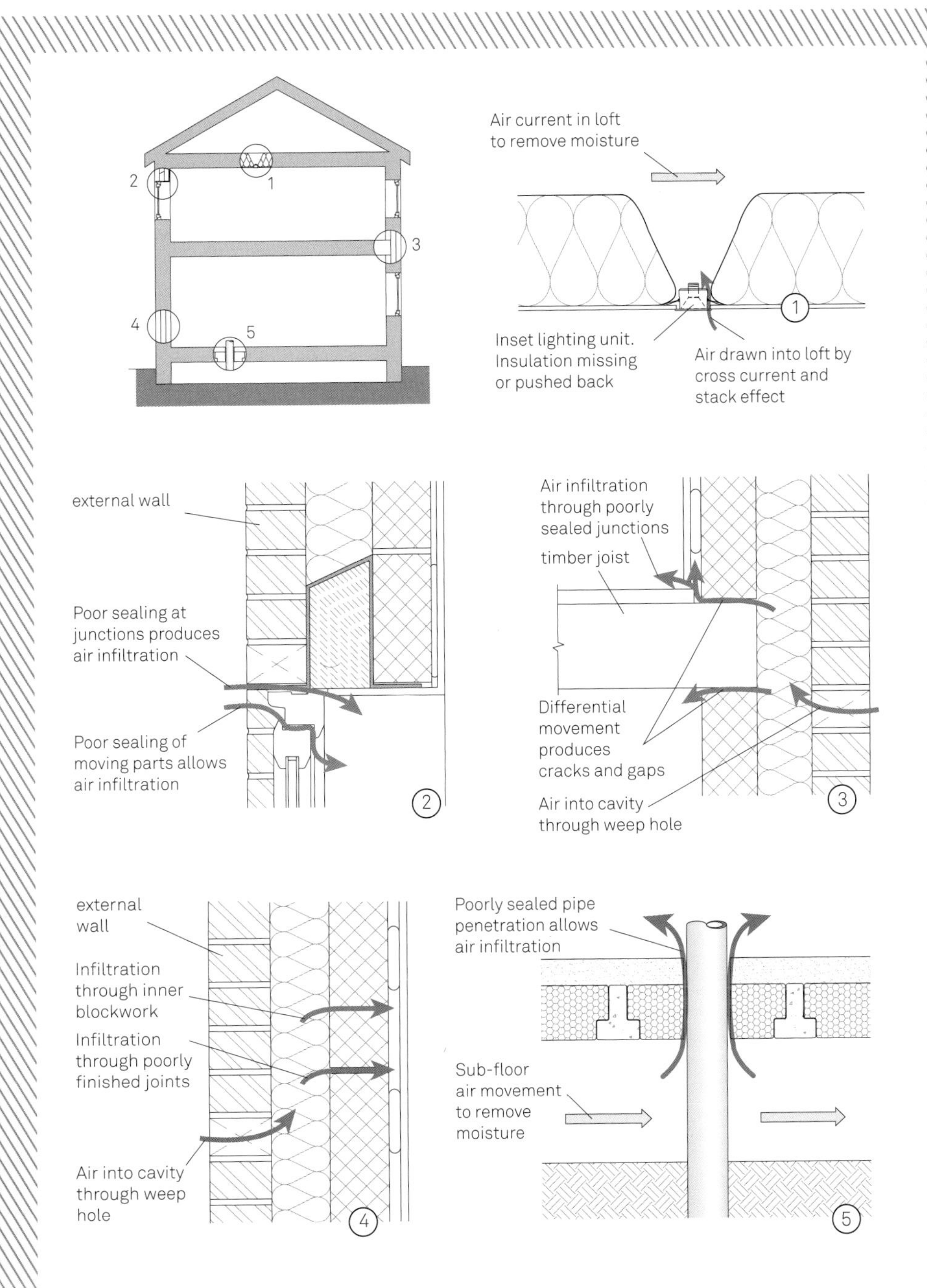

FIGURE 2-06 TYPICAL ROUTES FOR AIR INFILTRATION

Assessing infiltration

The susceptibility a building to air infiltration is expressed by its **air permeability**, that is, the volume of air which passes through each square metre of its exposed surface in an hour with a pressure difference of 50 Pa between inside and outside. It is expressed as $m^3/m^2.h$ at 50 Pa (cubic metre per square metre per hour at fifty pascals).

The air permeability is measured by pressure testing, which is carried out using portable fans fitted to a doorway or other opening. The openings of the building are closed and all vents sealed, then the fans are used to raise the air pressure within the building. The rate of air flow generated by the fans is measured at several pressure differences, then the air flow at the standard pressure difference of 50 Pa can be calculated.

The overall rate of air exchange for a building or space is usually given in air changes per hour (ach). One air change per hour notionally represents the replacement of all the air in the space by an equal volume of incoming air. So, 2 ach for a space with a volume of 60 m^3 represents the movement of 120 m^3 of air. There is no simple relationship between the air permeability rate and ach, because one is based on the area of a building's envelope and the other on its internal volume.

The rate of air infiltration will depend on the air permeability of the fabric and the environmental conditions, including wind speed and direction, and the internal and external temperatures. Rates vary enormously between buildings: a house built to meet the Passivhaus standard[4] requires an air change rate less than 0.6 ach, while in the UK some existing dwellings have air change rates well over 20 ach.

Figure 2–07 shows the reduction in heating energy which can be achieved by reducing air infiltration for a number of building types.

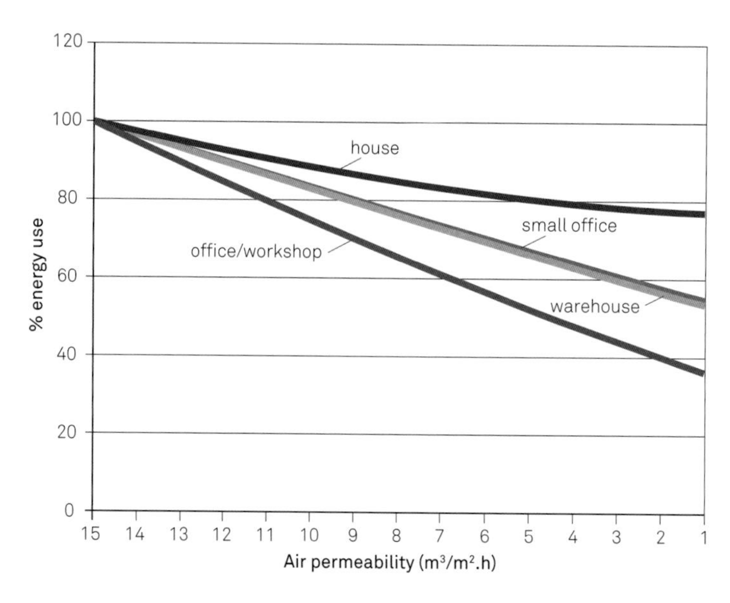

FIGURE 2-07 THE EFFECT OF INFILTRATION ON HEAT LOSS AND ENERGY USE
Data derived from SAP 2012 for dwellings and iSBEM for non-dwellings

Limiting infiltration

Minimising infiltration requires careful attention to the design and construction of building elements, junctions between elements and service penetrations:

* Lightweight aggregate blockwork is permeable and requires a parge coat of plaster to seal it. In contrast, structural insulated panels (SIPs) have low permeability because they have large, unbroken sheet facings bonded to insulation, with a limited number of junctions between panels.
* Cavities within construction should be avoided wherever possible, because introducing a cavity provides a route for air movement.[5]
* Junctions between elements should be designed without unnecessary gaps and cavities, and should be sealed, with tapes and sealants matched to the range of differential movement likely to occur. Laps are easier to seal than butt joints.
* Service penetrations offer significant scope for air leakage, as pipes, ducts and cables often run through elements. The optimum solution is to avoid penetrations wherever possible. Where that is not practicable, penetrations should be properly sealed, particularly at vapour control layers, air barriers and other membranes.
* Effective sealing of pipework and recessed light fittings is likely to require special components.

'BREATHING' BUILDINGS

It would seem logical to discuss 'breathing' buildings and 'breathing' constructions as part of broader consideration of air movement and infiltration; except that when people talk about 'breathing' in this context they are usually not talking about air movement, but about moisture transfer through the fabric by diffusion.

Breathing is a physiological process which involves air being drawn into an organism and then expelled. In contrast, a 'breather membrane' is a sheet of material with a low vapour resistance (i.e. defined in BS 5250:2011 as less than 0.5 MN/s) which, in the right conditions, facilitates the diffusion of water vapour: it will often be airtight. Clearly, a 'breather membrane' does not breathe: neither does a 'breathing wall'.

It is difficult to identify the source of this terminology, but one of the key references is a leaflet published by the Society for the Protection of Ancient Buildings (SPAB) in 1986, *The need for old buildings to 'breathe'*, which gives valuable guidance about problems with trapped moisture within traditional structures and the need to allow moisture transportation by diffusion and evaporation.

Frankly, the use of 'breathing' to refer to vapour transfer by diffusion is confusing and is unhelpful when we are attempting to clarify and disentangle the behaviour of air and moisture.

Air in the building fabric

While air infiltration through the building fabric should be avoided, there are situations where air currents are deliberately induced into air spaces to prevent damaging concentrations of moisture (see chapter 3: Moisture and the building fabric p. 83), for example:

* To the voids beneath suspended floors, in order to prevent ground moisture damaging the floor: this is particularly important for timber floors
* Behind impervious external layers in constructions, such as sheet metal covering laid on boarding
* In pitched roofs with vapour resistant underlay and a cold void above the thermal insulation

In each case, moisture moves into the cavity – either from the ground or from the building interior – potentially raising the moisture concentration to harmful levels. The air current replaces the moist air with external air, which should be drier. The air current will also increase the rate of heat loss through the element (see Figure 2–08):

* The provision of a current of external air effectively isolates the fabric beyond the cavity from the rest of the construction (i.e. the cavity surface acts as the external surface)
* The air current will increase infiltration by moving air through adjoining layers
* Where the layer adjacent to the cavity consists of fibrous insulation the pressure differences induced by the air current will result in air exchange between the insulation and the cavity

The simplest way to eliminate these effects is to adopt construction details which do not rely on air currents to avoid problems with moisture (see: chapter 3 Controlling interstitial condensation p. 88). Where that is not possible, for example in refurbishment or heritage projects, the cavity must be isolated from the building interior to limit heat loss.

Beneficial air movement in air spaces

There are three conditions where air currents within air spaces can be thermally beneficial:

* Dynamic insulation systems for cavity walls – consisting of rigid insulation boards containing vertical channels. Heat moving through the wall from the building interior warms the air within the channels, creating a stack uplift. The warmed air rises to a collector at the top of the wall to be fed directly into the occupied space or used in conjunction with a positive input ventilation system.
* Transpired solar collectors – consisting of micro-perforated cladding panels with a cavity behind them. The panels are warmed by solar gain, so air drawn through the perforations is also warmed. The air, which is further warmed by heat transferred from the building interior, rises up the cavity and is collected at the top and drawn into the building's ventilation system.
* Trombe walls – consisting of thermally massive walls faced with glazed cavities. Solar radiation passes through the glass and is absorbed by the wall, raising its temperature. Unventilated trombe walls rely on conduction to transfer the heat into the occupied space, while ventilated trombe walls have top and bottom vents which allow warmed air from the glazed cavity to reach the occupied space.

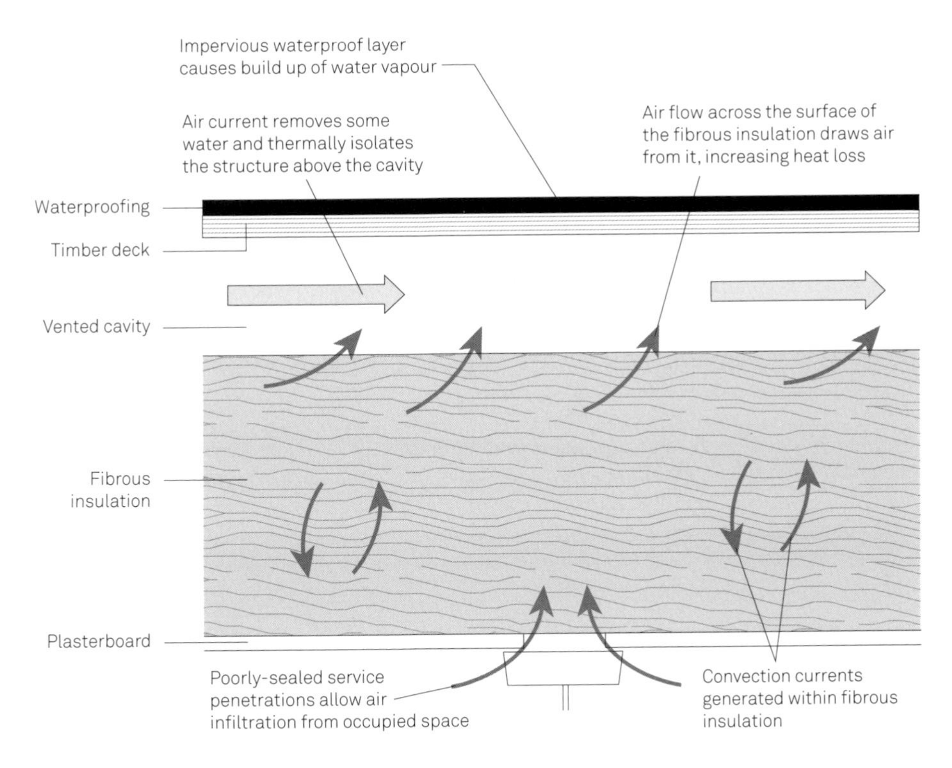

FIGURE 2-08 THE EFFECTS OF AIR CURRENTS IN CAVITIES ON HEAT LOSS IN COLD FLAT ROOF

Air and the building occupants

While the primary requirement for air within a building is that it maintains the health and wellbeing of the building occupants, air also affects the operation of some building services – such as combustion appliances used for heating – and has a role in moderating the internal temperature. In addition, in some geographic locations the prevention of the movement of radon gas from the ground to the occupied space is important in protecting building occupants from exposure to **radioactivity**.

This section examines those requirements for those four aspects of air supply, while the following section discusses ventilation methods.

Air for health and comfort

To protect the health and comfort of building occupants the air supply for a building must:

* Provide air which is suitable for respiration
* Remove pollutants (including tobacco smoke)
* Dilute odours
* Maintain moisture below nuisance levels

Respiration

The primary requirement of air for building occupants is that it is suitable for respiration, having the right levels of oxygen and carbon dioxide. The minimum concentration of oxygen for respiration is about 12% at normal atmospheric pressure, while carbon dioxide concentrations should not exceed 0.5%. Expired air contains approximately 16% oxygen and 4% carbon dioxide, so the main challenge when designing a building's air supply is to provide sufficient fresh air to keep the carbon dioxide concentration below the maximum exposure level, and preferably below 0.25%.

The air supply required to maintain those levels varies with the metabolic rate of the building occupants, which is, in turn, proportional to the activity level. Table 2.1 gives typical values. The rate of air flow provided by a ventilation system is usually quoted in litres per second per person (l/s per person). The ventilation rate required for a building or space is typically given in ach (air changes per hour). The air change rate can be combined with the volume of the building to define the rate at which air must be delivered by the ventilation system.

Providing suitable air change rates may mean that different ventilation strategies are needed for different parts of the same building.

TABLE 2.1 AIR SUPPLY RATES TO LIMIT CARBON DIOXIDE LEVELS

ACTIVITY	METABOLIC RATE (W)	FLOW RATE (l/s PER PERSON) TO MAINTAIN CO_2 LEVEL BELOW:	
		0.5%	0.25%
Seated quietly	100	0.8	1.8
Light work	160-320	1.3 - 2.6	2.8 - 5.6
Moderate work	320-480	2.6 - 3.9	5.6 - 8.4
Heavy work	480-650	3.9 - 5.3	8.4 - 11.4
Very heavy work	650-800	5.3 - 6.4	11.4 - 14.0

SOURCE: BS 5925:1991, TABLE 1.

Pollution

While pollutants such as gases, vapours, dust, fibres and biologically viable material (spores and bacteria) are present in the atmosphere to a varying extent, there are many activities within buildings which can produce pollutants in potentially harmful concentrations. Industrial and manufacturing processes can produce a wide range of pollutants, ranging from dust in sawmills to the emission of volatile chemicals in paintshops; office photocopiers and laser printers emit ozone and volatile organic compounds; underground car parks have high rates of carbon monoxide emission. Table 2.2 gives the UK's exposure limits for the commonest pollutants.

TABLE 2.2 MAXIMUM PERMITTED CONCENTRATIONS OF KEY POLLUTANTS		
POLLUTANT	TIME PERIOD	CONCENTRATION
Nitrogen oxide (NO_2)	1 hour average	288 µg/m³ (150ppb)
	annual mean	40 µg/m³ (21ppb)
Carbon monoxide (CO)	15 minute average	100 mg/m³ (90ppm)
	30 minutes average	60 mg/m³ (50ppm)
	1 hour average	30 mg/m³ (25ppm)
	8 hour average	10 mg/m³ (10ppm)
Total volatile organic compounds	12 hour average	35 mg/m³ (30ppm)
Ozone (O_3)	–	100 µg/m³

SOURCE: BUILDING REGULATIONS APPROVED DOCUMENT F 2010, APPENDIX A.
ABBREVIATIONS: PPB = PARTS PER BILLION, PPM = PARTS PER MILLION.

Potentially harmful concentrations of pollutants in buildings can be addressed by:

* Removing or controlling the source of pollution (e.g. changing a production process)
* Extracting the pollutant at the point of generation (e.g. fitting an extract hood to a cooker or an industrial saw)
* Filtration to remove the pollutant from the air (e.g. filtration of incoming air where there is heavy traffic outside the building)
* Reducing the concentration of the pollutant to safe levels by diluting it with incoming air (e.g. providing a large supply of fresh air in a workshop where volatile organic compounds are given off during manufacturing)

Odours

Human beings are sensitive to a wide range of airborne odours; some pleasant, some unpleasant. Within buildings people are subject to many odours, including body odour, toilet odour and cooking odour, with the mix largely dependent on the function of the building. The perception of odours, together with their intensity and quality – and whether or not they are acceptable – rests with individuals and cannot be measured directly.

Moreover, the perception of an odour – pleasant or unpleasant – changes over time. Sensitivity lessens over the first half an hour of exposure; over weeks and months it may come to be regarded as normal. The ventilation rates required to address unpleasant odours are therefore based on the perception of someone coming into the space from outside.

For spaces where the occupants are mainly sedentary a rate of 10 l/s per person will be sufficient (presuming the space to be a no-smoking zone).

Smoking

Tobacco smoke requires specific consideration, because it is both unpleasant and harmful: it causes irritation of mucous membranes in the respiratory system and eyes, while longer term exposure increases susceptibility to respiratory problems and, of course, lung cancer.

Air supply levels to counteract tobacco smoke have been based on considerations of odour for a non-smoker entering a space where smoking is permitted. A rate of 15 l/s per person is recommended for a space with typical number of smokers, rising to 40 l/s where all occupants are assumed to be smokers. The ventilation rates required to avoid mucous membrane irritation are approximately a quarter to a half of the rates for odour control.

Bans on smoking in public places and work places means that the need to provide ventilation for smoking is not as significant as 20 or 30 years ago.[6]

Moisture levels

High moisture levels in buildings can adversely affect comfort, induce problems that have an adverse impact on health, such as mould growth and house dust mites, and damage the building fabric (see chapter 3: Moisture and the building fabric p. 83). Water vapour forms part of the air within a building, so the main mechanism by which moisture travels is air movement.

Moisture levels may be controlled by drawing moisture-laden air out of the building as close as possible to the point of generation (e.g. in bathrooms, or laundry/cooking areas), while introducing drier air from elsewhere in the building. The necessary rate of air flow will depend on the rate of moisture generation (see chapter 3: Human activity p. 80).

Setting air supply rates

The consideration of each aspect of air supply for the health and comfort of building occupants results in a range of requirements for minimum rate of air flow: the requirements for respiration being the lowest and the requirements for the control of odours being the highest. In practice, providing the air supply necessary for the control of odours will meet most of the other requirements. The exceptions, where a higher air supply rate may be required, are:

* Spaces where pollutants are generated
* Spaces where moisture is generated

Air for combustion

Combustion appliances, such as heating boilers and cookers, require a supply of oxygen for combustion and their exhaust gases have to be removed.

The air supply for combustion will depend on the type, size and fuel of the appliance, but is likely to be in the range of 0.4–30 l/s for each kilowatt of rated output. For appliances which draw combustion air from the room, the ventilation requirement is commonly expressed in terms of a **free area of vent** (the total unobstructed opening area of the vent), or the **equivalent area** (the area of a single opening which would provide the same air flow as the vent, taking into account the effect of air turbulence at vent openings). An insufficient air supply can lead to harmful levels of carbon monoxide (defined in table 2.2).

For appliances with flues or chimneys, the exhaust gases (including carbon monoxide and carbon dioxide) will be drawn up the flue and expelled as a result of the stack effect. No further ventilation provision is required. The exhaust gases from unflued appliance, such as gas cookers and gas heaters, must be removed or diluted by ventilation to maintain concentrations below hazardous levels. The main requirement is to keep carbon dioxide levels in the space below 0.5%.

Balanced flue appliances, which draw air directly from outside and expel exhaust gases outside, such as domestic gas boilers, have no impact on the air supply to the rest of the building, although their exhaust flues, like any other, must be sited away from air intake vents.

Air for moderating temperature

Air movement has a role in moderating the internal temperature of buildings. When the internal temperature rises above the external temperature – as a result of solar gain, **metabolic gains** and gains from appliances – the warmed air within the building may be extracted and replaced by a supply of cooler air. Depending on the ventilation strategy (see below: Ventilation p. 60) the cooler air may be fresh, external air or conditioned air which has been cooled and perhaps de-humidified. The effectiveness of such cooling depends on the air change rate (which will be constrained by the need to avoid discomfort from air currents (see below: Air currents p. 68), the volume of the space and the relative temperatures of the internal surfaces and outside air.

Air movement can also be used during the night to absorb the heat that has been stored in thermally massive structures and PCMs during the day. Drawing cooler external air through the building, and bringing it into contact with those parts of the structure with higher admittances will transfer heat from the fabric to the air, leaving the fabric able to absorb heat again in the course of the following day.

The rate of air flow provided for cooling the building is likely to exceed that required to maintain the health and wellbeing of building occupants.

Protecting building occupants from radioactivity

Radioactivity, or **radioactive decay**, is the process in which atoms of unstable isotopes give off charged particles and decay to another element. The charged particles are referred to as **ionising radiation** because they carry enough energy to strip electrons from atoms they encounter, leaving the atoms with a positive electric charge. Radioactivity is measured in **becquerel** (Bq): 1 Bq is defined as one transformation or decay per second.

Ionising radiation damages human tissue. For instance, alpha particles emitted during radioactive decay are highly ionising and particularly harmful when ingested or inhaled. They damage lung cells by creating free radicals and damage DNA, which can produce cancerous mutations.

Although any exposure to ionising radiation brings an increased risk of harmful effects, the probability of harm from long-term, low-level exposure is proportionate to the extent of exposure.

The main natural source of radioactivity on earth is radon, a gas formed as part of the decay chain of uranium-234. Radon-222 is carcinogenic when inhaled: it decays in the lungs, emitting alpha particles, and the decay products, which include polonium, are deposited on lung tissue, where they decay further. Radon is the second highest cause of lung cancer in the UK, resulting in 1,100 deaths per year.[7]

Radon gas occurs naturally in the ground below buildings. It is formed in the geological strata, and rises through faults in the rock to reach the ground surface. It decays to other elements within a few days, so the concentration of radon diminishes rapidly from the source of generation.

The typical outdoor concentration of radioactivity from radon is 10–20 Bq/m^3.

Radon rising under a building will collect and penetrate the fabric. The air pressure in buildings is usually slightly lower than that outside, so the under-pressure will draw radon into the building. Also, radon is heavier than air, and the relatively sheltered conditions inside the building can result in high concentrations of radon. Figure 2–09 shows typical routes for radon penetration. Radon is colourless and odourless so the building occupants will not be able to detect it.

FIGURE 2-09
ROUTES OF RADON PENETRATION INTO BUILDINGS
Based on BR 211:2007, Figure 1

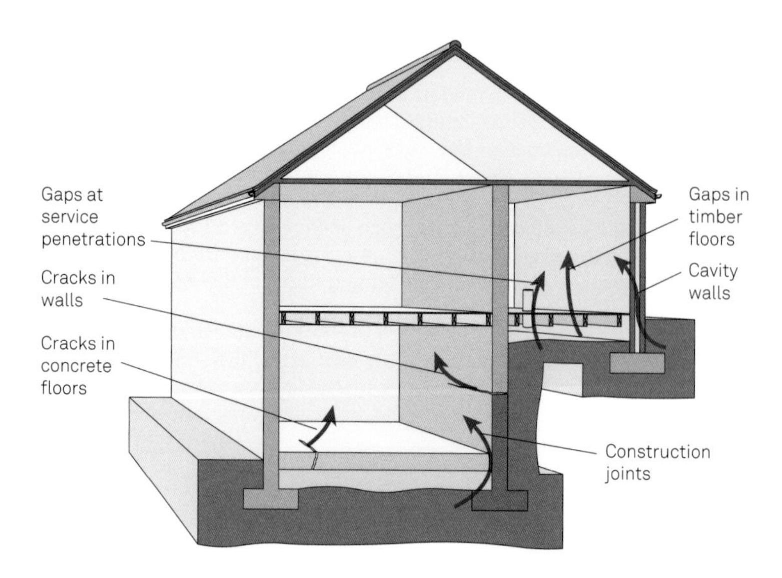

Although any exposure to radon brings a risk of harm, an indoor radon concentration of 200 Bq/m³ is recognised by the World Health Organization as the level at which radon protection measures are necessary in dwellings. In much of the UK the action level of 200 Bq/m³ is accompanied by a target level of 100 Bq/m³, at which radon protection should be considered.[8]

For new buildings, the radon concentration cannot be measured directly at the design stage, so the need for protective measures is established by consulting radon maps, which in the UK are prepared by the British Geological Survey. The maps show those areas where there is a significant likelihood of a dwelling having a radon concentration above the action level. Radon is particularly prevalent in areas with higher concentrations of uranium, such as those with granite-based geology (e.g. the Lake District and Cornwall).

Radon concentrations in existing buildings can be established by measurement. The standard method is to use passive detectors to record the amount of radioactivity taking place over a three to six month period.

Radon protection measures are commonly divided into passive measures, used at lower risk levels, and active measures used at higher risk level.

New buildings – passive protection measures

Passive measures limit the transmission of radon from the ground to the occupied space. They commonly involve the installation of a radon-resistant membrane, typically low-density polyethylene (LDPE), across the entire footprint of the building as far as the outside face of the external walls. (This is sometimes referred to as a radon barrier.) The construction details will vary, but groundbearing concrete floors, such as that shown in Figure 2–10, can incorporate a gas-resistant membrane between the slab or deck and the screed.

New buildings – active protection measures

Where higher concentrations of radon are likely, ventilation is required below the radon-resistant membrane to draw radon from beneath the building and reduce the pressure of radon against the underside of the membrane.

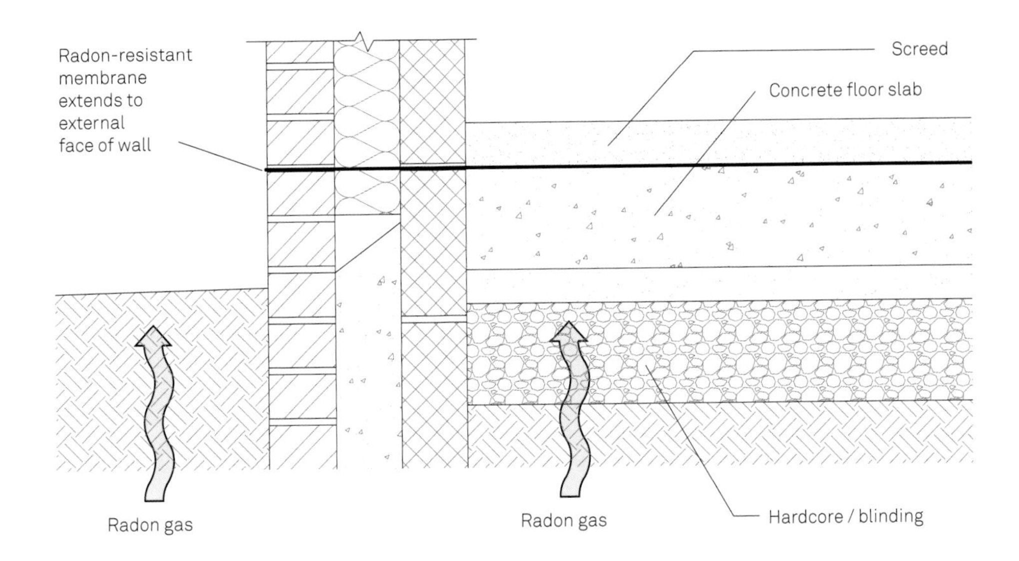

FIGURE 2-10 PASSIVE RADON PROTECTION MEASURES ON A GROUNDBEARING CONCRETE FLOOR

Providing ventilation beneath a suspended floor is straightforward, as shown in Figure 2–11. Groundbearing floors will require collection chambers, referred to as radon sumps, which are vented to the atmosphere: pre-fabricated plastic sumps are commonly used.

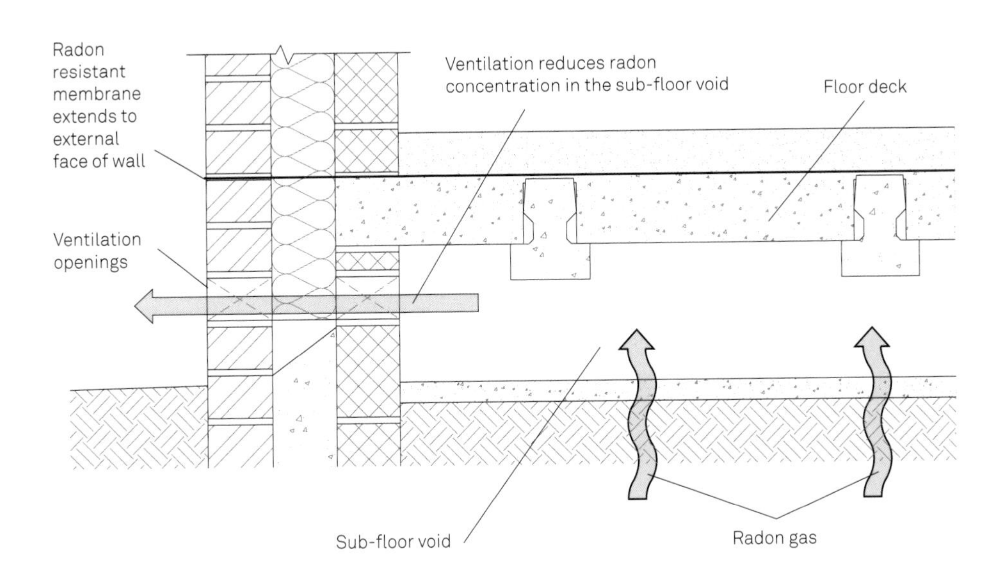

FIGURE 2-11 ACTIVE RADON PROTECTION MEASURES FOR A SUSPENDED FLOOR

Active systems should include provision for retro-fitting of an extraction system (mechanical or passive stack) which can be installed if high radon concentrations were to be found inside the building after construction.

Existing buildings

It is rarely possible to fit a radon-resistant membrane to an existing building. Remedial measures are therefore designed to:

* Reduce radon pressure beneath the floor – by fitting a radon sump under a solid floor or by increasing the amount of ventilation beneath a suspended floor. A radon sump can be formed from the outside by making a hole in the external wall below ground level, digging out a small cavity beneath the floor, then fitting an extract pipe which discharges at roof level. The best results are obtained with fan-powered systems.
* Prevent radon reaching the building interior – by installing positive pressure ventilation which forces air into the building and eliminates the normal under-pressure which draws radon into the building. Sealing gaps and cracks in floors can also reduce radon transmission, but may result in the decay of timber floors;
* Remove radon from the interior – by increasing ventilation rates. This will only be of limited benefit.

Ventilation

Having considered the factors that determine the air supply requirements, we can now consider how to provide those requirements – which are unlikely to be the identical in all areas of a building, or even at all times. A ventilation strategy will usually address four main functions:

* Background ventilation – providing the basic level of ventilation required to dilute odours and pollution, and maintain carbon dioxide levels within acceptable limits.
* Extraction – permanent or intermittent extraction of pollutants or high levels of water vapour. Extracting fumes and pollutants as close as possible to their source prevents their dispersion throughout the building.
* Purge ventilation – providing rapid removal of occasional pollutants. This can typically be provided by an openable window. It can also assist in addressing overheating by exchanging large volumes of air.
* Cooling – preventing overheating by replacing warmer inside air with cooler outside air when temperatures are high, or using night-time cooling to release heat stored in the thermal mass of the building during the day (see chapter 1: Overheating through openings p. 35).

Those functions can be address by three ventilation strategies:

* Natural ventilation, which relies on wind forces and the stack effect
* Mechanical ventilation, which uses powered fans to move air in and out of the building
* Mixed-mode ventilation, which augments natural forces with mechanical ventilation

In some cases, mechanical ventilation will be a necessity:

* Where toxic or noxious pollutants have to be removed at the point of release (typically this will apply to industrial processes)
* In some areas of hospitals, to prevent cross infection
* Where external air conditions would result in unacceptable quantities of pollutants being drawn into the building (e.g. if a building is adjacent to a busy road)
* Garages, car parks and other locations where exhaust gases and petrol vapour cannot be allowed to build up

In other cases mechanical or mixed-mode ventilation will be desirable, for example:

* To remove pollutants from factories and industrial processes
* To remove odours and moisture from dwellings
* In locations where a high density of occupation expected, such as lecture theatres

Finally, where natural ventilation is not inherently precluded, there are four additional considerations which may require mechanical or mixed-mode ventilation:

* Air quality – If either the supply or exhaust air needs to be filtered or cleaned because of outside air pollution, internal cleanliness requirements (e.g. clean rooms) or polluting processes within the building it is likely that natural ventilation will not be suitable. The flow-inducing pressure differences are so low that the additional resistance introduced by filters would seriously limit the amount of air movement
* Heat recovery – Heat losses resulting from ventilation air exchange can be limited by the use heat recovery units. However, heat recovery is incompatible with natural ventilation because the wind and stack forces driving natural ventilation will not be strong enough to overcome the resistance to air flow in the heat recovery unit
* Consistency – Given its inherent variability, natural ventilation cannot offer consistent flow rates, so natural ventilation is unlikely to be suitable for projects which require regular, predictable flow rates
* Control – Natural ventilation is unlikely to be suitable where design flow rates are variable (e.g. changing in number of occupants). Similarly, if the flow has to be in a given direction (e.g. always out of clean areas) wind-driven ventilation, with its changes in air flow, is unlikely to be suitable

For larger buildings, computer modelling will be required to design an effective natural ventilation system and to demonstrate that it will provide the required air supply.

Natural ventilation

Natural ventilation systems rely on wind forces and the stack effect to exchange air between the exterior and interior of a building and to move air within it. The three basic methods are:

* Cross-ventilation (Figure 2–12a)
* Single-sided ventilation (Figure 2–12b)
* Stack ventilation (Figure 2–12c)

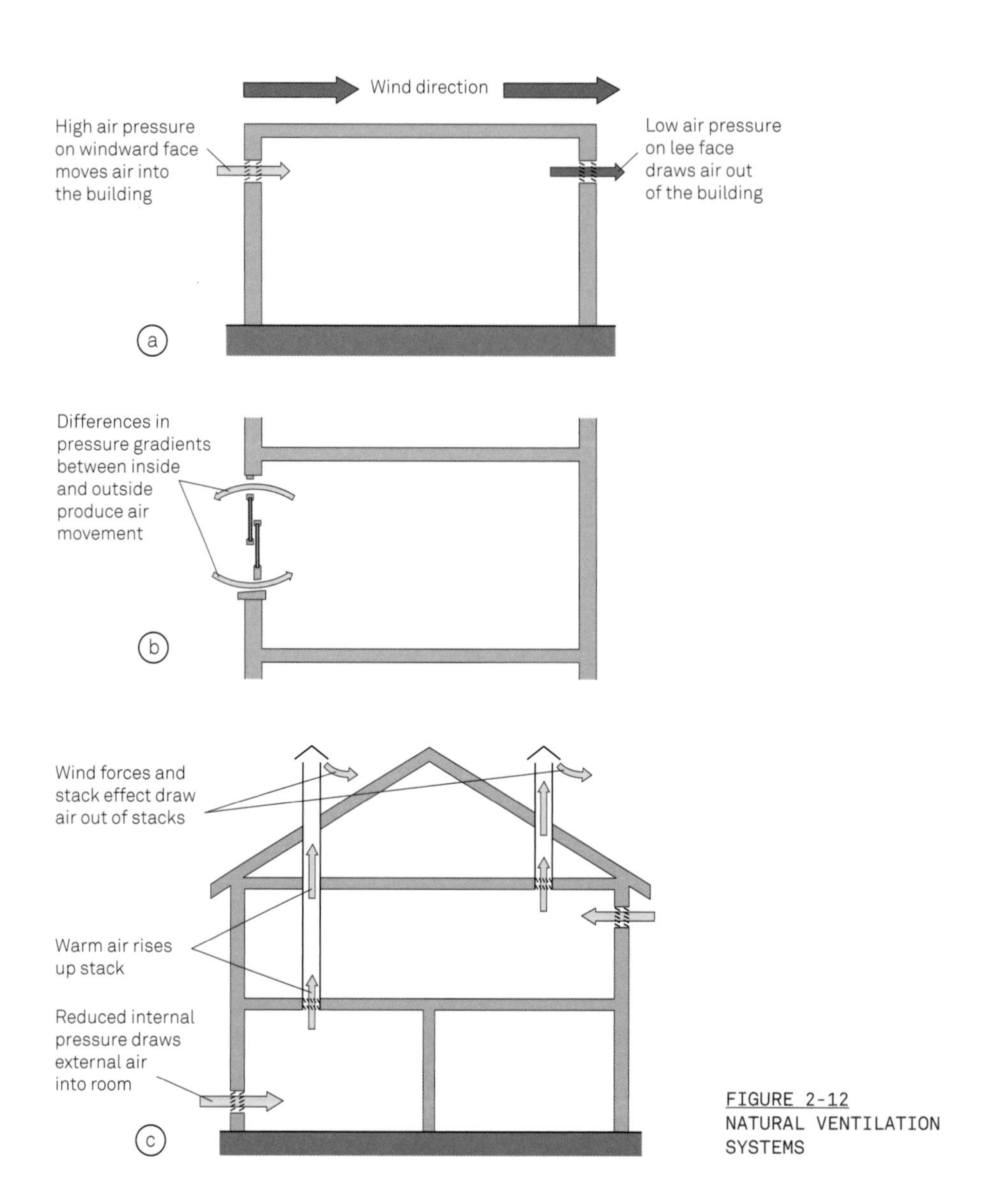

FIGURE 2-12
NATURAL VENTILATION SYSTEMS

Several strategies may be used on the same building: for example, the background ventilation in a house may be provided by trickle vents built into window frames, while extraction in the bathroom is provided by a passive stack vent running from the ceiling through the roof. Natural ventilation systems for large or complex buildings will utilise wind and stack forces together to provide air movement, for example, to ventilate a building with an atrium.

The wind and stack forces acting on the building will vary over the course of a year, even where there is a prevailing wind direction. As a result, not only will the amount of air exchange vary, but a vent may act as an inlet with some wind directions but draw air out of the building with others (and even when the wind is in one direction air currents will still eddy around the vents).

The design of the ventilation system must ensure that it will work satisfactorily under all conditions. In particular, incoming air must be clean, so vents must be sited away from localised sources of pollution, such as heating system exhausts.

Cross-ventilation

Cross-ventilation (Figure 2–12) relies on differential pressure generated across the building by the wind: air enters the building on the high-pressure windward side and is drawn out on the low-pressure leeward side. The rate of air exchange depends on the wind speed and direction, and the size and location of vents. As a rule of thumb, cross-ventilation will be effective where the depth of the space does not exceed five times the floor-to-ceiling height.

Typically, cross-ventilation to provide background ventilation will be provided by trickle vents above openings or incorporated into window and door frames, which will give about 0.1–0.2 ach. Open windows on opposing sides of a building will also provide cross-ventilation, and are often used to address overheating in summer. The necessary window area is usually specified as a fraction of the floor area, with 5% of floor area appropriate for sash windows or hinged lights opening more than 30 degrees, and 10% for those opening between 15 and 30 degrees. Fully open windows can provide 6–8 ach.

The direction of the prevailing wind should be taken into account; however, even when the wind is blowing parallel to the ventilation openings pressure differences across the building will still produce some air movement. Where ventilation openings are positioned at different heights (e.g. on different storeys) there will be some drive from the stack effect, which may reinforce or counteract the wind forces. Effective cross-ventilation requires a good flow of air between different sides of the building, so will be less effective in buildings where the interior is divided by many walls or partitions.

Single-sided ventilation

Rooms and spaces with openings to only one side, or with limited opportunity for cross-ventilation, can be naturally ventilated by exploiting the air movement generated by temperature differences between the inside and outside of the building (see above: The stack effect p. 46). Air movement can take place through twin wall vents where one is near floor level and the other is close to the ceiling, or, to a more limited extent, at the top and bottom of a large open window. (A large traditional sash window with top and bottom sashes opened will give a similar effect to twin vents.)

In the summer, when the internal air is cooler, air will be drawn in at high level and drawn out at low level. In the winter, when external air is cooler, air will be drawn in at low level and drawn out at high level (see Figure 2–05). The ventilation rate achieved by single-sided ventilation depends on the temperature difference and the distance between the upper and lower openings. There will also be some air exchange driven by the wind. On some faces of a building the wind force may be the predominant driver, in which case the ventilation rate will be determined by the wind speed.

Single-sided ventilation will be effective where the room depth does not exceed twice the room height. Trickle vents can provide 0.1 ach, while fully openable windows (sized to the rules given above) will provide 4–6 ach.

Stack ventilation

Stack ventilation (also referred to as 'passive stack ventilation') uses the stack effect to move air up and out of a building at high level, drawing fresh air into the building at lower level (Figure 2–12c). The temperature differential between air inside the building and outside the building results in the upward drive (see above: The stack effect p. 46), and wind action across the ventilation terminal also draws air out. The stack effect has been exploited in this way for thousands of years (see box: The wind tower p. 68).

Stack ventilation can be used as part of a whole building natural ventilation system, either using shafts to ventilate buildings several floors high, or simply using multistorey spaces such as atria as the stacks. Figure 2–13 shows how a central atrium can be used to ventilate the floors to either side of it, venting air from the top and drawing air into each floor. The performance of such systems depends on stack height, temperature difference and the free area of vents. Typically, the opening at the base of the stack should be 3–4% of the floor area.

Stack ventilation is often used instead of extract fans to remove moist air from bathrooms and other wet rooms, through vertical ducts which terminate above the roof. The rate of extraction is variable, because it depends on the temperature differential, which will be less in summer, and the wind speed.

Controls

Natural ventilation relies on climate-driven forces so the ventilation rates achieved will be variable. Controlling such systems is, in most cases, a matter of opening and closing vents. For dwellings and buildings of similar complexity, manual operation of vents and windows is generally sufficient; although where moisture control is the main function of the ventilation system, humidity controlled vents may be used. In larger buildings, controls can be integrated into the building management system, but are still limited to opening and closing vents, albeit in a sophisticated way.

Mechanical ventilation

Mechanical ventilation systems use fans to move air into, out of and around buildings. They range in complexity from simple extract systems – such as a cooker hood or shower room fan – to centralised, ducted, air-conditioning systems which incorporate filtration, heating and cooling of air and heat exchange. Mechanical ventilation systems allow substantially better control than natural ventilation systems, but they require power to operate, and need ongoing maintenance to maintain their efficiency.

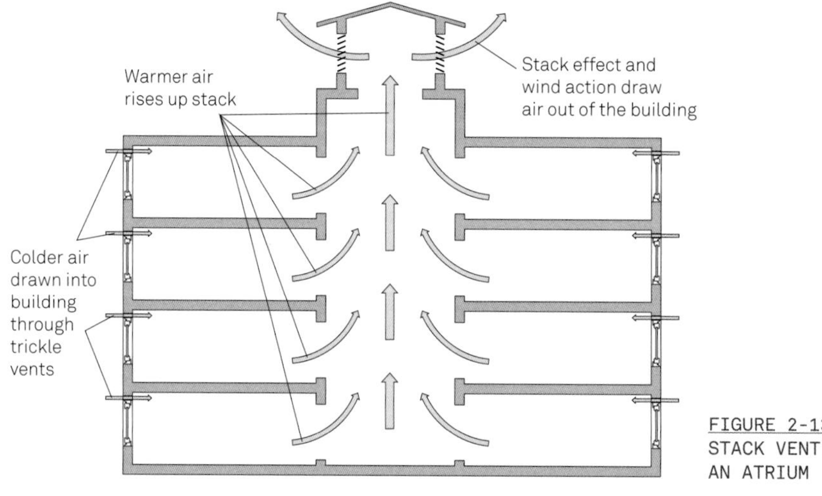

FIGURE 2-13
STACK VENTILATION IN AN ATRIUM

THE WIND TOWER

The Iranian *badgir*, or wind tower, is a traditional ventilation system, dating back to the fourth millennium BCE, which uses wind and stack effects to cool a building's interior.

The *badgir* consists of a tower, rising above the roof of the building, which is divided into a number of shafts (commonly four, but one, two and eight shafts are found) that run down into the building. Towers can be square or rectangular in plan, with the latter having the widest side perpendicular to the prevailing wind.

When the wind blows the windward shaft acts as an air inlet, while leeward shafts act as air outlets. A combination of wind forces and the stack effect results in warm internal air being replaced by cooler outside air. In a more sophisticated system (illustrated below), the incoming air is drawn along an underground water channel (*qanat*) which cools the incoming air by evaporation before it reaches the occupied space (see chapter 3: Vapour and heat p. 76).

The *badgir* is part of a broader temperature control strategy which minimises solar gain by the use of narrow streets, tall walls and small windows, and uses thermally massive fabric to make best use of the diurnal temperature swings.

The same fundamental principles can be employed in naturally ventilated buildings today.

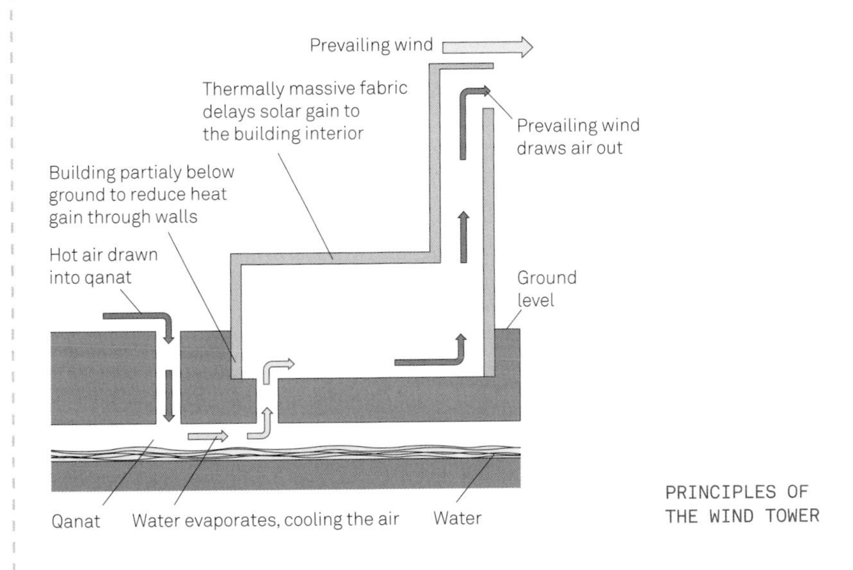

PRINCIPLES OF THE WIND TOWER

The three main types of mechanical system are illustrated in Figure 2–14 and described below.

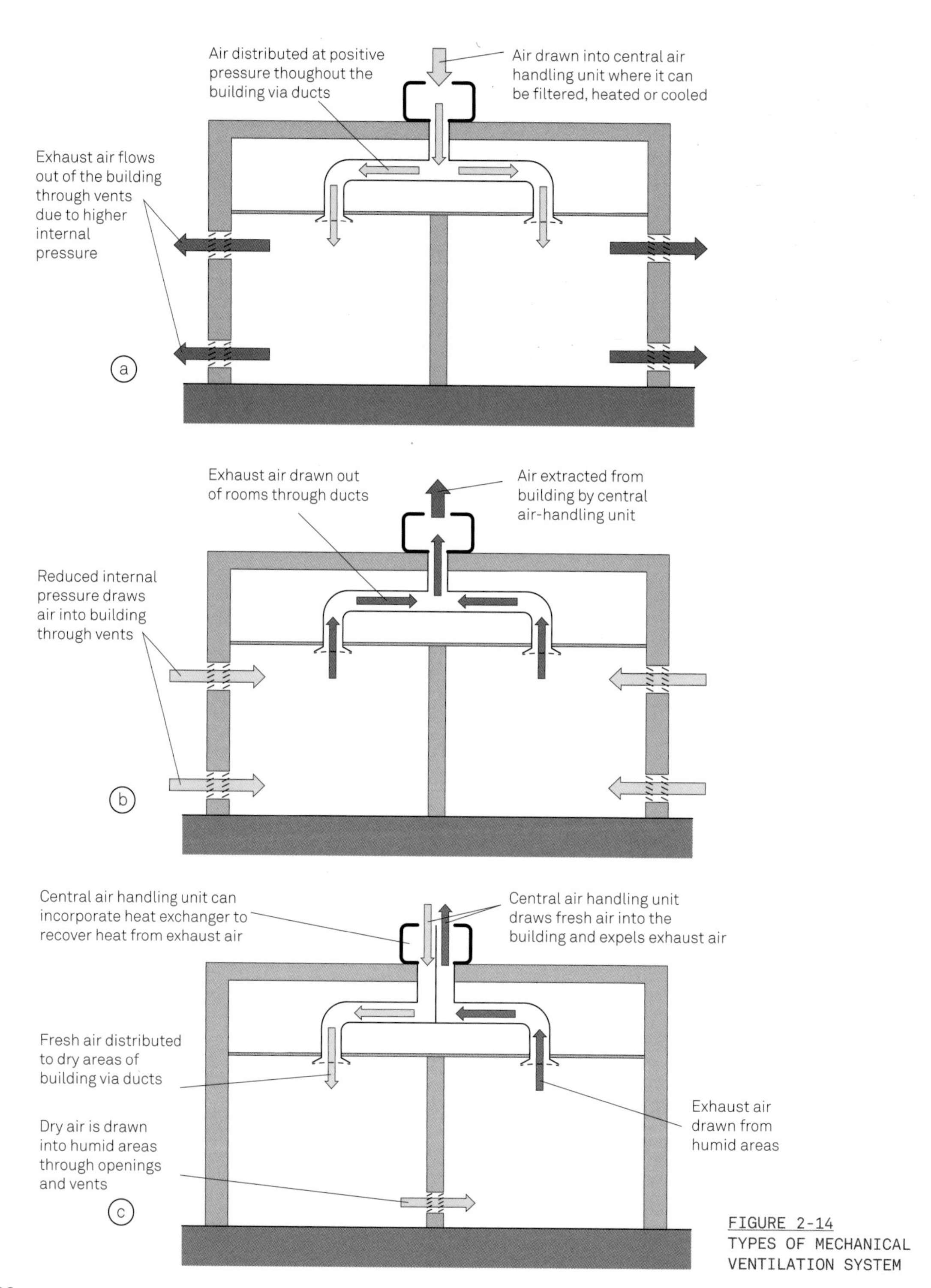

FIGURE 2-14
TYPES OF MECHANICAL VENTILATION SYSTEM

Supply-only ventilation

In a supply-only ventilation system (Figure 2–14a) air is drawn into the building to a central air-handling unit and circulated by ducts to different spaces in the building. The building is maintained at a higher air pressure than outside (the systems are often described as 'positive pressure systems'), so exhaust air is expelled through passive vents. Air can be filtered in the central air handling unit to remove contaminants before it is circulated, and it can also be heated or cooled.

Basic supply-only systems are often used in dwellings, while more complex systems which include filtration are used for operating theatres and clean rooms because the positive pressure prevents contamination from the rest of the building, and allows air to be filtered and conditioned before it is circulated.

Extract-only

In extract-only ventilation systems (Figure 2–14b) air is drawn out of the building or room either by centralised fans and ducts, or by individual fans. As the air pressure in the building is then lower than atmospheric pressure (commonly referred to as 'negative pressure systems') fresh air is drawn into the building through vents, louvres and other permanent openings.

Extract-only ventilation is well suited to addressing moisture levels, because such systems extract moist air at the site of generation, preventing it from travelling through the rest of the building. The extraction system can include filtration, which enables pollutants to be removed from extracted air before the air passes into the atmosphere, so preventing the further spread of pollutants.

Balanced ventilation

In balanced ventilation systems (Figure 2–14c) air is drawn into one or more central air handling units then distributed throughout the building by ducts to air outlets. Simultaneously, air is extracted from other outlets to the air-handling unit and passed out of the building. The use of separate input and output ductwork means that moist or polluted air can be extracted close to the point of contamination, which helps to prevent distribution through the building, and allows fresh air to be provided to specific locations

Balanced ventilation systems will often include a heat exchanger, to minimise the heat loss resulting from extraction and will often form part of full conditioning systems. The possibility of heat exchange makes balanced ventilation popular in dwellings with low energy use targets.

Mixed-mode ventilation

Mixed-mode ventilation is a hybrid approach which integrates natural ventilation (through openable windows and vents) with mechanical ventilation systems. Mixed-mode ventilation operates in three main ways:

* Complementary – The mechanical ventilation runs at the same time as the natural ventilation is working. Often, the mechanical system provides the background ventilation, while building occupants can open windows if they want to. Alternatively, the mechanical system may be used to augment natural ventilation when weather conditions result in low flow rates.
* Changeover – The ventilation provision changes between natural and mechanical according to need. The changeover may be seasonal, with windows openable in mild weather but locked shut in winter; or daily, with natural ventilation during the day and mechanical ventilation at night to provide cooling.

* Zoned – Some spaces in the building are naturally ventilated while others are mechanically ventilated. For example, open-plan office areas may be naturally ventilated, while meeting rooms – which have higher ventilation requirements – are mechanically ventilated.

Mixed-mode ventilation can reduce energy consumption and allow smaller mechanical systems to be installed. It also improves the building occupants' satisfaction by giving them more control over their conditions. However, it is more complex to specify and requires careful commissioning to ensure natural and mechanical systems do not work against each other.[9]

Design considerations

Regardless of whether natural, mechanical or mixed-mode ventilation is adopted there are several common design considerations: effectiveness, air currents, noise and heat loss.

Ventilation effectiveness

The effectiveness of a ventilation system measures how well it delivers the supply air to the building's occupants. A system which extracted the supply air before it reached the occupants would have an effectiveness of 0 (this can happen if a supply and an extract grille are located too close together), while a system which mixed the supply air fully with the room air before it was breathed by the occupants would have an effectiveness of 1. Displacement ventilation, which introduces cold air close to floor level and extracts warmed exhaust air at ceiling level, has an effectiveness greater than 1.

Air currents

Ventilation air introduced into a building will not be at room temperature (although it will not necessarily be at the temperature of outside air). Vents and inlets should therefore be sited so as to prevent building occupants being exposed to draughts of cold air. However, even if the incoming air is well mixed, air currents can, of themselves, affect the perceived temperature: currents of air travelling at 0.1 m/s can be felt by the building occupants and will lower the perceived temperature of the air. At air speeds greater than 0.15 m/s the temperature of the air should be raised to counteract the perceived cooling. Air speeds greater than 0.3 m/s will cause discomfort and are unlikely to be acceptable, except for summer cooling.

Noise

A ventilation system which provides a direct connection for air movement between the interior and exterior of a building will also provide a route for sound transmission. Where external noise levels are high – for example because of traffic – sound transmission through vents and ducts may be disturbing to the building occupants.

For some projects the problem may be avoided by the use of acoustic vents which will attenuate the noise (see chapter 4: Controlling noise from outside p. 113), but in some locations the likely disturbance may be so great that a ventilation system which forms a direct connection between inside and outside will be unacceptable.

Ductwork can also transmit sound between rooms and zones, causing a nuisance and, in some cases, threatening privacy. The noise from the fans that are part of a mechanical ventilation system can also be transmitted through out a building via ventilation ductwork. Attenuating ductwork or silencers may be fitted to reduce noise levels.

Heat loss

The deliberate movement of air into and out of a building for ventilation purposes will result in heat transfer, which may result in a heat gain to the building (typically in the summer), or heat loss (mainly during winter).

To reduce the contribution of ventilation to heat transfer, mechanical ventilation systems can include heat exchangers, which transfer heat from warm exhaust air to the colder fresher air which is being drawn into the building. Heat exchangers are rarely compatible with natural ventilation, because they have too great an effect on the air flow.

Air in the bigger picture

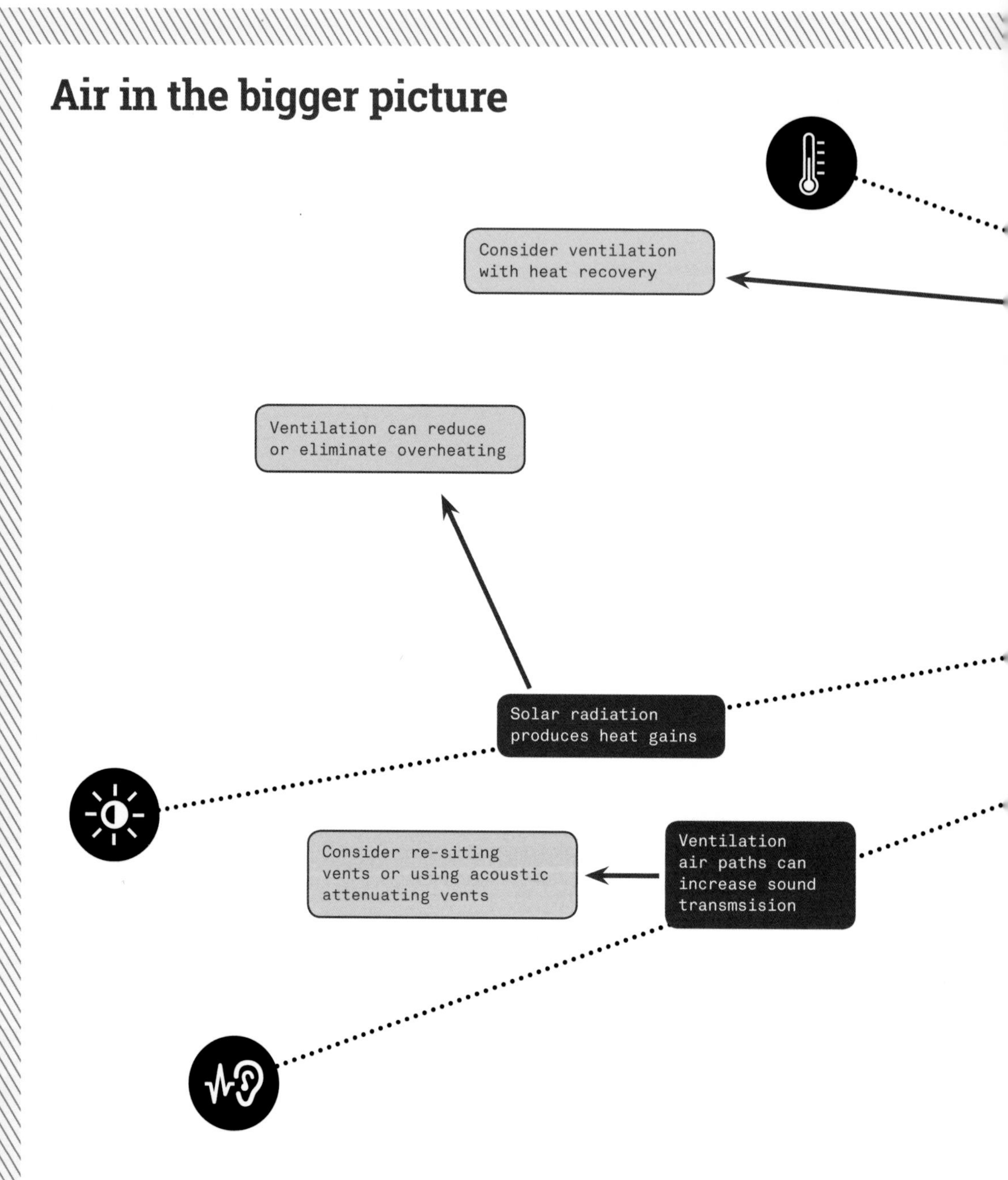

INTERACTIONS WITH AIR

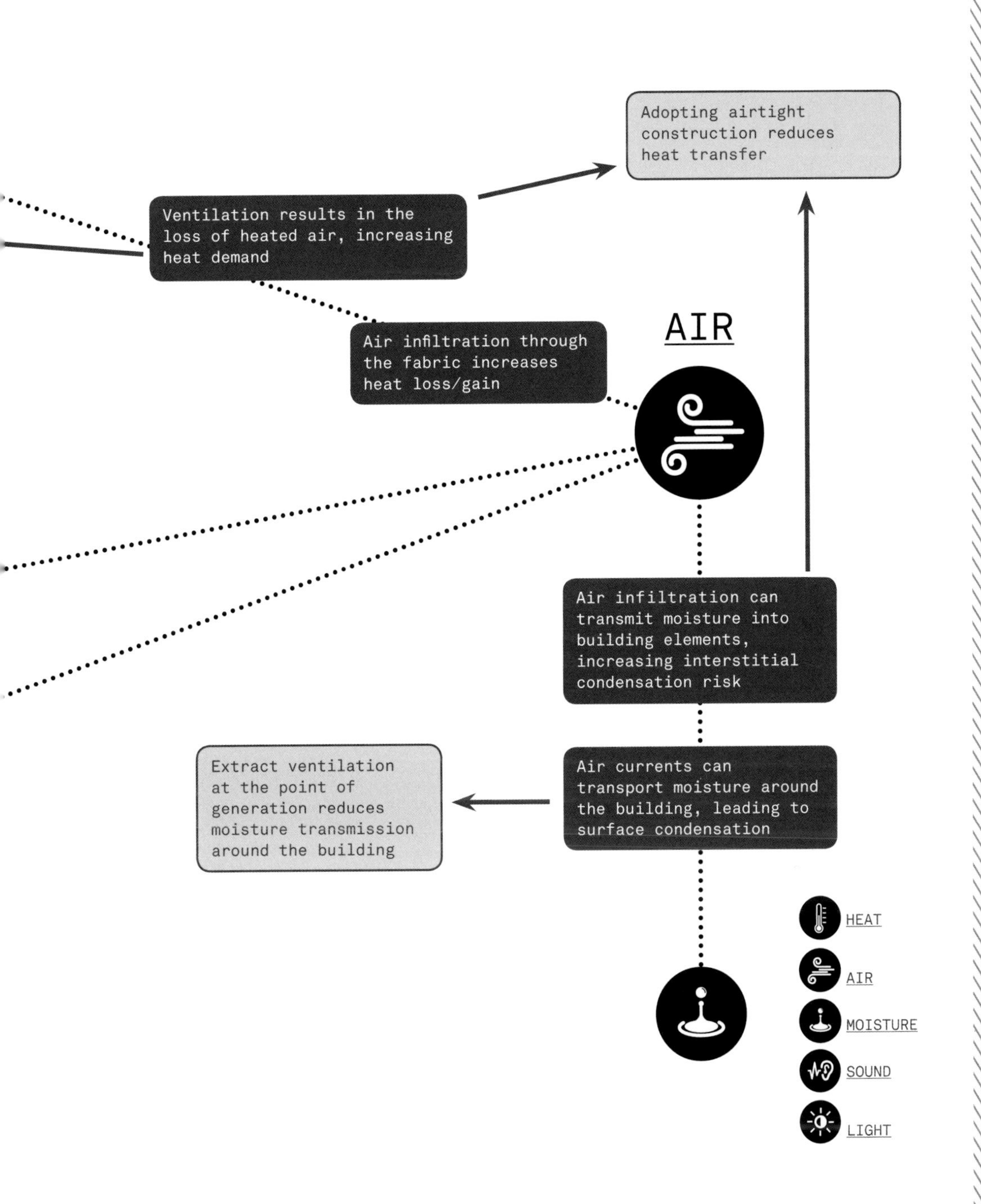

Adopting airtight construction reduces heat transfer
Ventilation results in the loss of heated air, increasing heat demand
AIR
Air infiltration through the fabric increases heat loss/gain
Air infiltration can transmit moisture into building elements, increasing interstitial condensation risk
Extract ventilation at the point of generation reduces moisture transmission around the building
Air currents can transport moisture around the building, leading to surface condensation
HEAT
AIR
MOISTURE
SOUND
LIGHT

03

Moisture

<u>DEFINTION:</u>
MOISTURE: WATER IN VAPOUR FORM, WHICH IS PRESENT IN AIR AND CAN DIFFUSE THROUGH AND CONDENSE ON SOLIDS.

The relationship between humans and water is intimate and complex. As well as drinking water to survive, we also release it: we breathe, with exhaled air having a higher water content than inhaled air; and we sweat, as part of the physiological regulation of body temperature. And beyond the animal, there are the activities of living, commerce and culture which use and release water into the atmosphere: we wash ourselves and our clothes, we cook, we use swimming pools.

Although we need a certain amount of moisture in the atmosphere for comfort, in buildings, too much can cause discomfort or problems such as surface and interstitial condensation, which in turn damage the building fabric and harm its occupants. We therefore need to design and operate buildings in ways which properly manage the movement of moisture.

This chapter will first consider how water vapour behaves – particularly when subject to temperature changes. It will then set out the moisture levels necessary for human comfort, before examining the problems caused by excessive levels of moisture, and how they may be prevented. At the same time we must bear in mind the strong interactions in the behaviour of moisture, air and heat.

The fundamentals of moisture

Water vapour is water in its gaseous phase. It is produced by the evaporation of liquid water as part of the continuous cycles of evaporation and condensation which occur within the earth's atmosphere. The atmosphere contains 1–2% of water vapour by volume, but the amount of water in a particular location will depend on the local climate.

In order to understand the problems that water vapour can cause in buildings, and how they may be prevented, we need first to understand the behaviour of water vapour, and how it may be measured.

Moisture content and vapour pressure

The amount of water vapour in a volume of air can be expressed as by its **moisture content**. This is the mass of water vapour for each kilogram of dry air – dry air being the atmospheric gases considered without water or water vapour – in an air/water vapour mixture, which is not the same as the amount of water vapour in each kilogram of air. Moisture content is usually expressed in g/kg (grams per kilogram): air at standard pressure at 20°C and 70% humidity has a moisture content of about 12 g/kg.

Like other gases, water vapour exerts pressure as its molecules collide with each other and with other materials (see chapter 2: The behaviour of gases p. 42). This is the **vapour pressure**, and is generally measured in kilopascals (kPa). Increasing or reducing the amount of water in the air increases or reduces the vapour pressure.

The choice of measurement – moisture content or vapour pressure – depends on the purpose of analysis. Vapour pressure is used for assessing condensation risk because differential vapour pressure drives vapour movement. Moisture content is more useful for design and sizing of air-conditioning systems, because it gives an immediate assessment of the quantities of moisture involved.

Saturation vapour pressure, relative humidity and percentage saturation

There is a limit to the amount of water vapour which can be present in an given amount of air. Air containing that maximum amount is said to be saturated. The amount of water vapour at saturation varies with temperature: warmer air contains more water vapour at saturation than cooler air.

Saturation is expressed in terms of the **saturation vapour pressure** (SVP) which is the vapour pressure exerted by saturated air. The SVP is directly affected by temperature: air at a higher temperature has a higher SVP than air at a lower temperature, as Figure 3–01 shows.

Closely related to the SVP is relative humidity, which expresses the vapour pressure as a percentage of the SVP. For example, if the vapour pressure of a body of air is 1.0 kPa and the SVP of that air is 1.4 kPa then the relative humidity is 71%. Relative humidity is usually expressed as % rh.

Because the SVP changes with temperature, the relative humidity will also change with temperature. Consider a volume of air which has a vapour pressure of 1 kPa (shown as the solid

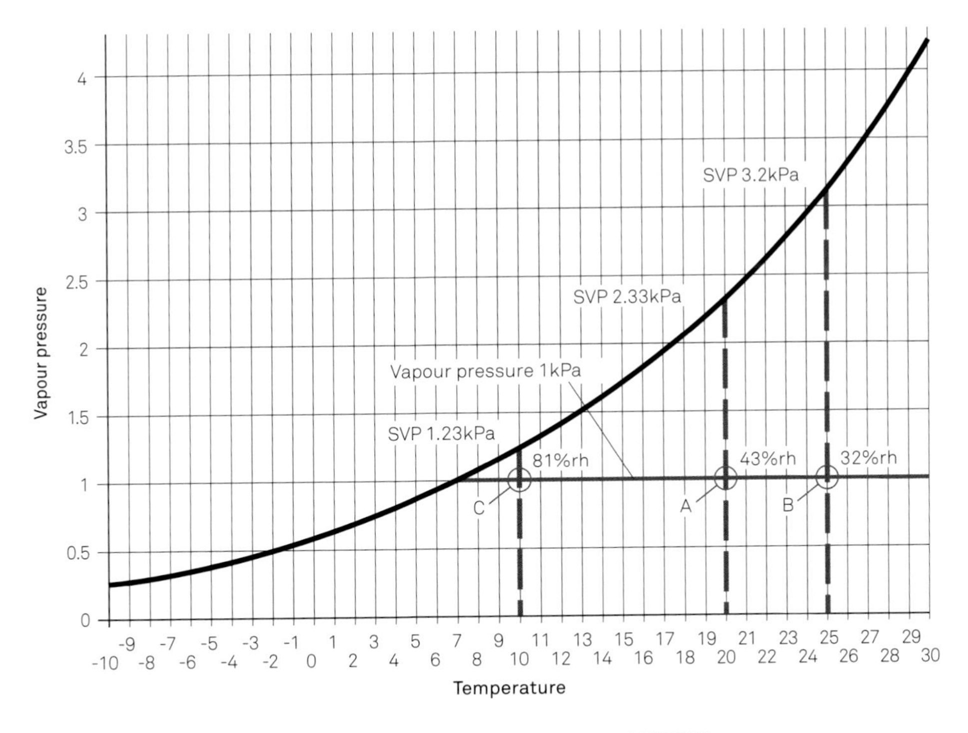

FIGURE 3-01 SATURATION VAPOUR PRESSURE AND RELATIVE HUMIDITY

horizontal line in Figure 3–01): the SVP of air at 20°C is 2.3 kPa, so at that temperature the air has a relative humidity of 43% rh (point A in Figure 3–01). If the air is heated to 25°C, the SVP rises to 3.2 kPa and the relative humidity drops to 32% rh (point B). Conversely, if the air is cooled to 10°C, the SVP drops to 1.23 kPa, which means the relative humidity rises to 81% rh (point C). Because relative humidity is linked to SVP, it can never exceed 100%.

The moisture content of air can also be expressed in terms of mass, and is the maximum weight of moisture in the air at that temperature: for example, the saturation moisture content of air at 20°C is 14.75 g/kg. The amount of water vapour in a sample of air can be expressed as a percentage of its saturation water vapour content: this is the **percentage saturation** (%): for example, air at 20°C with a moisture content of 7.8 g/kg would have a percentage saturation of 52.8%.

The relative humidity and percentage saturation values for air are identical at 0% and 100%, but diverge between those two points: in temperate climates the divergence is less than 1%, but at higher temperatures (40°C plus) the difference can be as much as 10%. As both measures are expressed in percentages (and the tell-tale 'rh' may well be missing) it is crucial to clarify which is being used on any project. Typically, the analysis of interstitial condensation risk will use relative humidity, while the design of building services will consider percentage saturation.

Vapour and heat

Evaporation – the change of state from liquid to vapour – requires energy, whereas condensation, the change from vapour to liquid, releases energy: this is the latent heat discussed in chapter 1 (see: Phase change materials p. 35). The total amount of energy in moist air is measured by its specific **enthalpy** (the total of its sensible and latent heat), which will vary as it is heated or cooled, or its moisture content changes. Enthalpy is used when calculating the energy involved in humidification and dehumidification.

The latent heat of evaporation is employed to determine the moisture content of air using paired **dry bulb** and **wet bulb** thermometers in a swing psychrometer. A swing psychrometer is so called because it has a dry bulb and a wet bulb thermometer mounted on a swivel handle, allowing them to be spun round quickly so they travel at more than 2 m/s. The dry bulb thermometer measures ambient air temperature, and the wet bulb thermometer, which is kept wet in muslin, is cooled as water evaporates into the air. The amount of water which can evaporate depends on the moisture content of the air. Drier air allows more water to evaporate before it reaches saturation, taking more heat and therefore giving a lower web bulb temperature. The vapour pressure can then be calculated from the dry and wet bulb temperatures using a standard equation.[1]

Dew point

The dew point is a way of considering saturation in terms of temperature. When a body of air is cooled its SVP will also be lower and, at a certain temperature, will be the same as the actual vapour pressure. This is referred to as the **dew point temperature**. If the air is cooled below its dew point temperature its new SVP will be lower than the original vapour pressure and the excess water vapour will condense. The formation of dew on the ground when warm, moisture-laden air cools under a clear night sky illustrates both the name and the phenomenon.

Dew point is useful for understanding the role of temperature in the behaviour of water vapour and highlights the fact that a lowering of temperature, without any other change, can result in condensation. It is particularly useful when analysing interstitial condensation risk (see: Controlling interstitial condensation p. 88).

The psychrometric chart

The most important design tool for understanding the behaviour of water is the **psychrometric chart**, which is a graphical expression of the relationships between the properties of water vapour in air. Figure 3–02 shows a slightly simplified version. The y-axis (A) shows the vapour pressure in kPa. For psychrometric charts intended for analysis of services, the amount of water vapour is expressed as the moisture content (g/kg). The x axis (B) shows the dry bulb temperature in degrees Celsius. The saturation line (C), is either 100% saturation or 100% relative humidity. Running at an angle to the x axis, with the scale following the saturation line, is the wet bulb temperature (D). Finally, Percentage saturation or relative humidity lines (E), are usually at 10% spacings.

A point can be fixed on the chart by using two pieces of information, allowing other values to be read off. So, if we know the dry and web bulb temperatures we can read off the moisture content, or the percentage saturation, or relative humidity. We can also use the chart to see what happens to water vapour in air when conditions change. For example, looking at Figure 3–03, we can use the temperature, vapour pressure and relative humidity lines to see what happens to water vapour in a body of air. Point A represents outside air with a vapour pressure of 0.75 kPa at a temperature of 5°C. The air has a relative humidity of 86%. At B the air has been heated to 20°C, with the moisture content remaining the same; the relative humidity is now 32%. By point C the vapour content of the air has increased to 1.4 kPa as a result of activity in the building; the relative humidity is now 60%. By point D the air has cooled to 15°C, with the same moisture content; the relative humidity is now 82%. Finally, by point E the air has cooled further to 12°C. The relative humidity is 100%.

Psychrometric charts can be used to understand the behaviour of water vapour travelling through building elements and also to determine the changes which need to be made to temperature and humidity of air in order to provide the desired internal climate.

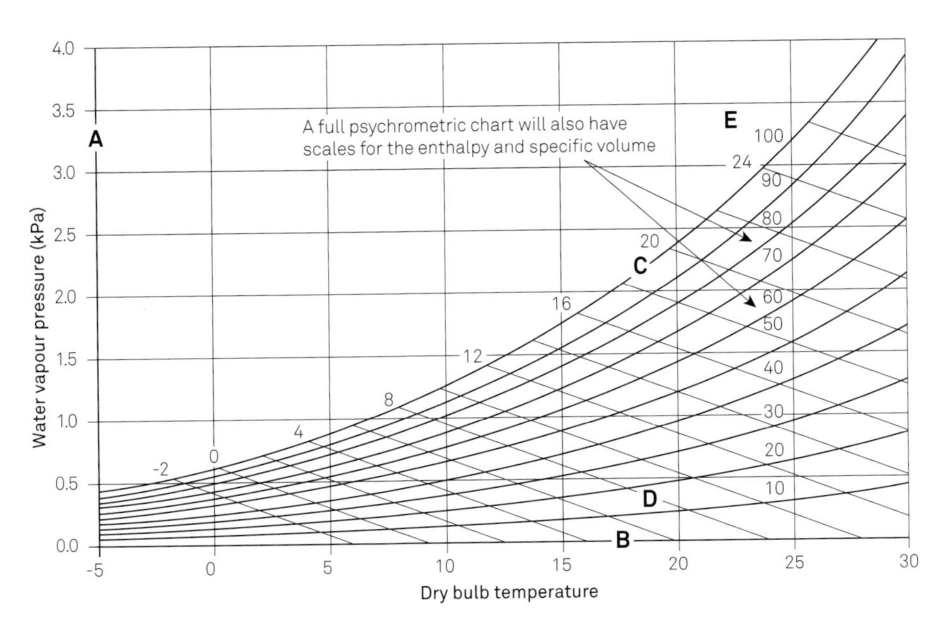

FIGURE 3-02 A PSYCHROMETRIC CHART
Based on BS 5250:2002 Figure A.2

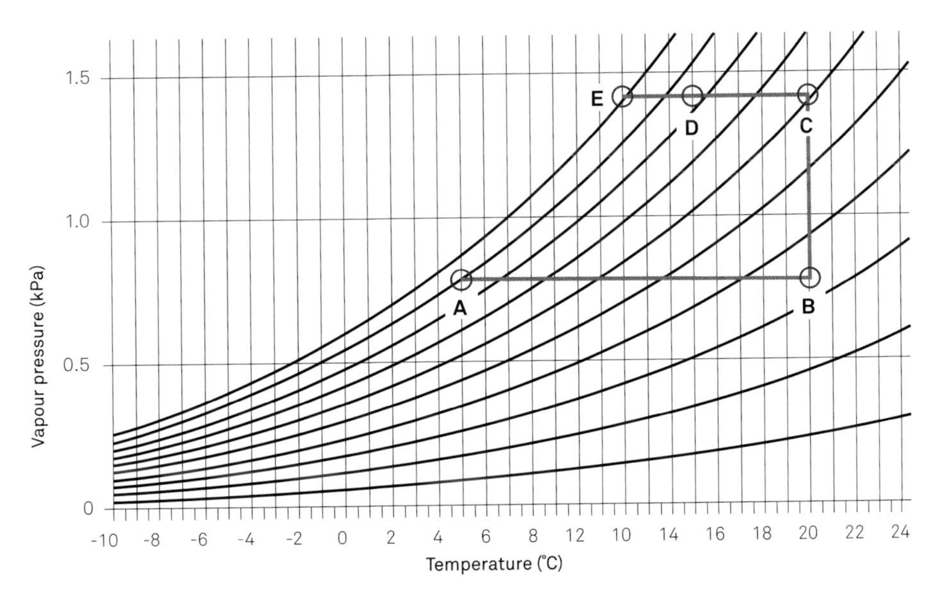

FIGURE 3-03 CHANGING CONDITIONS FOR WATER VAPOUR SHOWN ON A PSYCHROMETRIC CHART

How water vapour moves

In common with other gases, water vapour is continually in motion, being driven by two main mechanisms: bulk air movement and diffusion.

Bulk air movement

As air moves, so the water vapour in the air moves with it. Bringing air into a building for ventilation will also bring in the water vapour component of that air, while extracting air will also remove its water vapour. Air infiltration through the building fabric (see chapter 2: Infiltration p. 48) will also move water vapour into the fabric.

Diffusion

Water vapour tends to equalise its pressure by diffusing from high pressure to low pressure areas. The water vapour in a space with high vapour pressure – such as a bathroom or laundry – will diffuse to adjoining spaces where vapour pressures are lower, tending to equalise the pressure through all the spaces.

Water vapour can also diffuse through solid materials. A difference in vapour pressure across a material results in water vapour diffusing through the material from the side with the higher pressure to the side with the lower pressure. For example, water vapour will diffuse through the ceiling of a bathroom into a loft space above. The rate of diffusion is affected by the size of the pressure difference (greater differences result in a faster diffusion rate) and the vapour resistance of the material.

Capillary action

Although not strictly a mechanism of vapour movement, **capillary action** results in water moving within construction materials, and is worth briefly considering. Porous materials – including plaster, brick and timber – will absorb liquid water which can then move through the material, downward as a result of gravity, or sideways and upward as a result of capillary action.

Capillary action occurs in narrow spaces in the material: the hydrostatic forces within water (which result in its surface tension) and the adhesive forces between the liquid and the surrounding material act to draw the water further into the space, with narrower spaces producing a greater rise. The height to which water can rise depends on the sorptivity of the material (see below: Moisture penetration p. 83).

Vapour resistance

The diffusion of water vapour through solids can be expressed in two ways: vapour permeability or vapour resistance.

Vapour permeability is a measure of how rapidly water vapour diffuses through a material. It is expressed as the rate of transfer through one metre of the material and measured in gm/MNs (gram-metres per mega-newton second). Water vapour will diffuse more rapidly through materials with higher vapour permeability. The rate of diffusion through the specific thickness of a material in a building element is expressed by its **vapour permeance.** For example, if a material has a vapour permeability of 0.02 gm/MNs, then a 200 mm thickness will have a vapour permeance of 0.1 g/MNs.

Vapour resistivity is a measure of the resistance of a material to water vapour diffusion. It is expressed for one metre of the material and measured in MNs/gm (mega-newton seconds per gram metre). Water vapour will diffuse more rapidly through materials with lower vapour resistance. The resistance to water vapour diffusion of a specific thickness of a material in a building element is expressed by its **vapour resistance.** For example, if a material has a vapour resistance of 50 MNs/gm, then a 200 mm thickness will have a vapour resistance of 10 MNs/g.

Vapour permeability and vapour resistance are related; the vapour permeability of a material being the inverse of its vapour resistivity. The analysis of water vapour diffusion through building elements generally employs vapour resistance because it is analogous to the thermal resistance used in calculating heat transfer (see chapter 1: Heat transfer through solids p. 6), with the same relationship to thickness.

For most materials the vapour resistance is usually calculated from the vapour resistivity and thickness, but for membranes, such as vapour open membranes and vapour control layers, the vapour resistance is established by testing. Typical vapour resistivities and resistances are shown in tables 3.1 and 3.2.

TABLE 3.1 VAPOUR RESISTIVITIES OF COMMON MATERIALS

MATERIAL	TYPICAL VAPOUR RESISTIVITY (MNs/gm)
Mineral wool (rock wool, glass fibre)	5
Brick	50
Concrete block – medium weight	50
Plasterboard	60
Concrete – medium weight	150
Sheathing plywood	150–1000
Oriented strandboard	200–500
Polyisocyanurate	300
Expanded polystyrene (EPS)	300
Extruded polystyrene (XPS)	1000
Metals	Effectively infinite

SOURCE: BS 5250:2011 TABLE E.1.

TABLE 3.2 VAPOUR RESISTANCES OF COMMON MEMBRANES

MATERIAL	TYPICAL VAPOUR RESISTANCE (MNs/g)
Vapour open tiling underlay	0.2
Paint	0.5
Polyethylene (0.12 mm, 500 gauge)	250
Bituminous tiling underlay	450
Polyethylene (0.25 mm, 100 gauge)	500
Aluminium foil	1000
Foil-encapsulated membrane	10000

SOURCE: BS 5250:2011 TABLE E.2.

There is a huge variation in the vapour resistances of membranes, and it is important to differentiate between a membrane to be used as a vapour control layer (VCL) to prevent water vapour entering a building element and a 'vapour open membrane', which is designed to allow water vapour to leave a building element. The vapour resistance of a VCL can be as much as 100,000 times as great as that of a vapour open underlay. Selecting a suitable membrane requires consideration of environmental conditions and the position of the membrane within the construction (see below: Controlling interstitial condensation p. 88 for more detailed guidance).

European standards and product data sheets generally quote the water vapour resistance factor, μ, or equivalent air layer thickness, s_d. Those values can be converted to vapour resistivity and vapour resistance respectively by multiplying the values by five.[2] For example, a membrane with an s_d of 0.4 would have a vapour resistance of 2.0 MNs/g.

Sources of moisture in buildings

There is always moisture in buildings, moving in and out of the occupied spaces and the fabric by mass transport, diffusion and capillary action. This section identifies the four main sources of that moisture: the atmosphere, construction, accident and human activity.

Atmospheric moisture

The amount of water vapour in the open air varies with location, season and time of day. In naturally ventilated buildings the continual exchange of air between the inside and outside moves water vapour in and out of the building. Consequently, the background moisture content of internal air is close to that of outside air. In fully conditioned buildings the moisture content of internal air should be independent of the moisture content of the outside air.

Precipitation

Rain and snow falling on a building, or driven onto it by the wind, may be absorbed by the building fabric. Preventing the ingress of precipitation through the building fabric is a matter of competent design and construction, and as such, beyond the scope of this book.

Ground moisture

Moisture in the ground will be absorbed by porous construction materials. It can then move through the fabric and rise by capillary action.

Construction moisture

Concrete, masonry and plaster require large amounts of water during the construction process. Some of that water is locked permanently into the building fabric, but the remainder of the moisture evaporates, much of it into the building interior, raising the humidity level, particularly during the first year following completion. There may also be additional moisture resulting from precipitation during construction, although competent builders should work to reduce that to a minimum.

Accidental moisture

Moisture can arise from isolated events such as flooding and leaks (both through the fabric and from building services). Both have the potential for increasing moisture levels within a building. However, when they do occur the risk of condensation is probably a long way from the occupants' minds.

Human activity

Finally, moisture is given off by the occupants of a building (both human and animal) and generated by their activities. As discussed in chapter 2, humans give off moisture in breathing and sweating: average hourly values for an adult human vary from 40 g/hour when asleep to 300 g/hour when doing manual work. Domestic activities such as showering and bathing, laundry

and cooking generate substantial amounts of moisture (see table 3.3). Commercial, industrial and leisure activities (laundries, manufacturing and swimming pools) can also generate significant amounts of moisture.

TABLE 3.3 RATES OF MOISTURE GENERATION	
ACTIVITY	MOISTURE GENERATION RATES
People - asleep	40 g/h per person
People - seated, office work	70 g/h per person
People - standing, housework	90 g/h per person
People - moderate manual work	300 g/h per person
Cooking - electricity	2000 g/day
Cooking - gas	3000 g/day
Dishwashing	400 g/day
Bathing/washing	200 g/day per person
Showering (15 minutes)	600 g/shower
Washing clothes	500 g/day
Drying clothes indoors	1500 d/day
Washing floors	200 g/day
Plants	20 g/day per plant

SOURCE: REPRODUCED FROM BS 5250:2011, TABLE D.6.

Moisture and the building occupants

The amount of moisture in the air plays a crucial part in maintaining comfort conditions within buildings. For occupant comfort, humidity should be in the range 30–70% saturation. (Note that, when dealing with comfort conditions it is usual to work in percentage saturation, rather than relative humidity.) Drier air causes irritation and dryness of the eyes and nasal passages, while higher levels of humidity reduce the evaporation of perspiration and so limit the body's ability to regulate temperature properly.

The range of comfortable humidity and temperature (generally taken as dry bulb temperatures of 18–23°C) can be plotted on a psychometric chart to show a comfort envelope, as in Figure 3-04. (Here, the temperature is the 'operative temperature', which for still air is the average of the air and radiant temperatures.) Of course, other requirements, such as maintaining suitable conditions an industrial process, may be more important than maintaining standard comfort conditions.

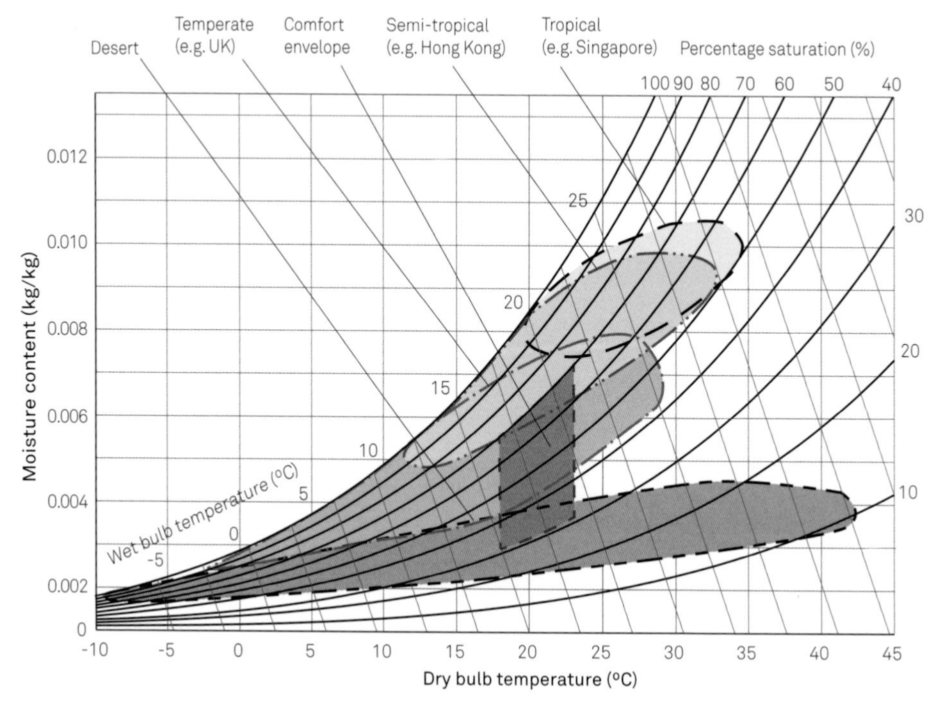

FIGURE 3-04 COMFORT CONDITIONS WITH ENVIRONMENTAL CONDITIONS IN DIFFERENT CLIMATE ZONES

Maintaining comfortable humidity in fully air conditioned buildings

For fully conditioned buildings, the temperature and humidity of the air within the building can be controlled using a combination of heating, cooling, humidification and de-humidification. The combination of those processes required for a building will depend on the external climate conditions.

The required changes in temperature and humidity can be determined using psychrometric charts, starting from the external conditions, and calculating the transformation needed to achieve the designed internal conditions. The conditioning equipment can then be selected and sized on the basis of those transformations. Seasonal variations in climate will need to be taken into account in the specification process.

Maintaining comfortable humidity in other buildings

In buildings which are only heated or cooled, the humidity levels will vary with the moisture content of external air, the rate of moisture generation within the building and the rate of air movement between the interior and exterior. That may result in buildings being too humid or to dry at some times of the year. While building users (and particularly those living in affected dwellings) will tend to adopt humidifiers or de-humidifiers, addressing moisture generation and ventilation rates is likely to be as effective, and more energy efficient.

Moisture and the building fabric

Although water vapour is required in buildings to maintain comfort conditions, significant problems can result from excessive moisture levels. The main issues are:

* Penetrating moisture
* Mould growth on surfaces
* House dust mites
* Surface condensation
* Interstitial condensation

The rest of this section addresses the first four issues and the typical techniques for addressing them. The last two, however, are more complex and merit separate sections of their own.

Moisture penetration

Rain and snow can penetrate roofs through poorly formed junctions, badly sized gutters or defects such as missing tiles. Deep snow lying on a roof may melt from beneath (as a result of heat loss from the roof): the resulting melt water may be trapped under the remaining snow and back up into the roof where it may cause damage to the fabric, or penetrate the interior of the building. This is a common problem for buildings where the gutters of pitched roofs are set behind parapets (e.g. traditional churches). The risk of water penetration can be reduced by installing heating wires or mats into the gutters, in order to maintain the drainage route for melt water.

Wind-driven rain and snow can also penetrate walls through inadequately protected junctions: the moisture can also be absorbed by brickwork and mortar and travel through cracks, or be drawn by capillary action to the internal face of the wall to cause staining or mould growth. If moisture-laden masonry is exposed to freezing temperatures the expansion of water can result in spalling. These problems may be avoided by specifying the fabric with regard for the local climate conditions.

Ground moisture rising by capillary action – commonly referred to as rising damp – can cause structural damage to timbers (particularly floors), damage internal finishes and cause mould growth.

Preventative measures, such as damp-proof membranes and damp-proof courses, will prevent ground moisture becoming a problem in new buildings. Although rising damp does cause problems in existing buildings, the majority of problems turn out to be the result of condensation, poor drainage or water ponding against the building fabric.[3]

Mould growth

Mould growth on building surfaces or within materials can cause health problems, notably respiratory problems, and exacerbate existing conditions such as asthma. Mould also damages building fabric, fittings, furniture and belongings: it is also unsightly, and its presence can be distressing to building occupants.

Mould spores are widely present in the air, and will germinate and grow when there is warmth, oxygen, nutrients and moisture. As most buildings have sufficient oxygen, are warm enough, and have suitable nutrients, the limiting condition is the moisture level. Moulds are hydroscopic and do not require liquid water for germination: a relative humidity above 80% is sufficient for germination, while they can continue to grow at lower humidities.

Mould growth can be prevented by ensuring relative humidity levels at internal surfaces do not regularly exceed 80%. In a heating season, the relative humidity at surfaces will generally be 10% higher than that of the air, consequently, the relative humidity of the air will need to be less than 70% to avoid mould growth.

The recommendations for preventing surface condensation (discussed later in this chapter) are equally applicable for preventing mould growth.

House dust mites

House dust mites (*Dermatophagoides pteronyssinus*) are creatures about 0.25 mm in length which live off organic detritus, such as flakes of human skin. The mites themselves, together with digestive enzymes in their droppings (which can persist long after a mite has died), can trigger and exacerbate allergic conditions including asthma and eczema. Mites do not drink, but absorb atmospheric moisture, so flourish in the moist, warm conditions present in many dwellings.

Much of the guidance on reducing problems caused by house dust mites addresses removing mites and their droppings from bedding and soft furnishings. However, a significant contribution to control can be made by bringing moisture levels below 60%rh to limit their reproduction.

Controlling surface condensation

Surface condensation is one of the commonest problems caused by high moisture levels: it results in:

* Nuisance dripping
* Damage to finishes
* Structural damage, particularly to window frames and surrounds
* High humidity levels in hygroscopic materials which can lead to decay

How and where surface condensation forms

Surface condensation occurs when warm, moist air comes into contact with a colder surface and is cooled below its dew point. The SVP at the new, lower temperature is less than the air's original vapour pressure: the excess moisture is unable to remain as water vapour so its state changes to liquid, and it is deposited on a nearby surface. In buildings, this process is manifested as the steamed-up mirror in the bathroom, or water droplets on a window or pooled on the window frame of a winter bedroom.

The formation of surface condensation can be seen graphically in Figure 3–05: air at a temperature of 20°C, with a vapour pressure of 1.25, has a relative humidity of 53%. When it comes into contact with a cold surface it is cooled to 10°C. It is now below the dew point temperature

of 10.2°C, so becomes saturated and the excess moisture condenses. The amount of condensate deposited depends on the vapour pressure and the air and surface temperatures; it is also affected by the rate at which warmer air passes the surface.

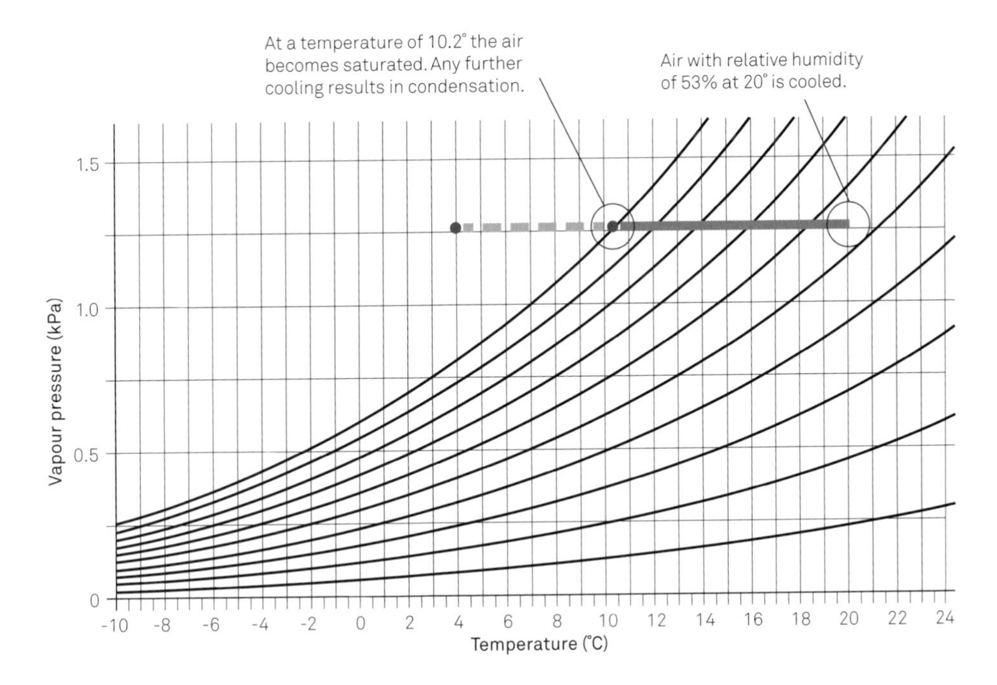

FIGURE 3-05 FORMATION OF SURFACE CONDENSATION

Typically, surface condensation occurs at the internal faces of poorly insulated external elements (mainly in existing buildings) where high rates of heat loss result in low surface temperatures (see Figure 3–06).

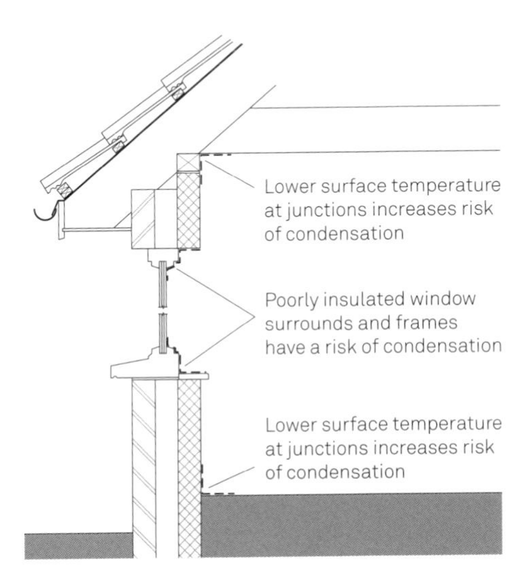

FIGURE 3-06
LOCATIONS WITH A RISK OF SURFACE CONDENSATION

In new buildings with good standards of insulation for floors, walls and roofs there is little risk of surface condensation on elements themselves, but there is a risk at junctions between elements where the rates of heat loss are higher, and temperatures are consequently lower (see chapter 1: Heat transfer at junctions p. 16). For example, the surface temperature of a poorly designed window reveal with insufficient insulation may well be low enough for surface condensation to take place. The wall–floor junction can also be cold enough for surface condensation.

Localised cold spots can also be formed by metal fixings and ties which penetrate insulation layers, resulting in pattern staining on the wall, as surface condensation and or mould growth are seen.

Condensation also forms on glazing – particularly single glazing – where it will tend to run, pooling on frames and sometimes on window boards. It can also form on uninsulated cold water pipework and cisterns which are cooled by the water within them.

Finally, surface condensation can be a significant issue in unheated and uninsulated buildings, such as metal-clad agricultural buildings used for housing animals. Animals give off substantially more water vapour than humans and in winter the warm, moist air within an animal shed air comes into contact with the metal cladding which – as a result of night sky radiation – will often be very cold, leading to large amounts of condensate forming on the underside of the sheeting.

Preventing surface condensation

To prevent surface condensation we must address moisture levels within a building and the temperatures of internal surfaces.

Controlling moisture levels

Moisture levels can be controlled by well-designed ventilation which extracts moist air as close as possible to the source of generation – to prevent moisture travelling to other parts of the building – and replaces it with drier air.

Localised extraction systems based on intermittent fans or vents may be used to draw moist air from the building (see Figure 3–07a). However, extraction will increase energy use, because the replacement air will need to be heated or cooled. It may be beneficial to use balanced ventilation systems which recover heat from humid exhaust air and use it to warm incoming replacement air fed into drier spaces (see Figure 3–07b and chapter 2: Ventilation p. 60). Air flows at the room level must be considered to prevent a build-up of moisture in poorly ventilated 'dead' areas. (A classic example is a large wardrobe put against an external wall: mould growth behind it is almost guaranteed.)

Controlling surface temperatures

Cold surfaces of building elements result from high rates of fabric heat loss. For heated buildings, the thermal performance required to meet new-build energy efficiency standards will generally be sufficient to avoid cold surfaces, provided heating is adequate. (For example, guidance to the Building Regulations in England recommends the U-value of a wall be no worse than 0.70 W/m^2K to avoid condensation, but the starting design standard for energy efficiency requires a U-value of around 0.18 W/m^2K.) The main risk in such buildings comes at junctions between elements, where geometry and discontinuity of insulation result in increased heat loss.

The key measure of a junction for avoiding surface condensation is the **temperature factor**, which expresses the temperature drop at a surface (see box: Temperature factor p. 88). The basic principles of detailing (see chapter 1: Heat transfer at junctions p. 16) will ensure a suitable temperature factor. Even given high levels of insulation for elements and junctions, the temperature of a surface depends on the heat system being powerful enough to maintain the internal temperature of the building.

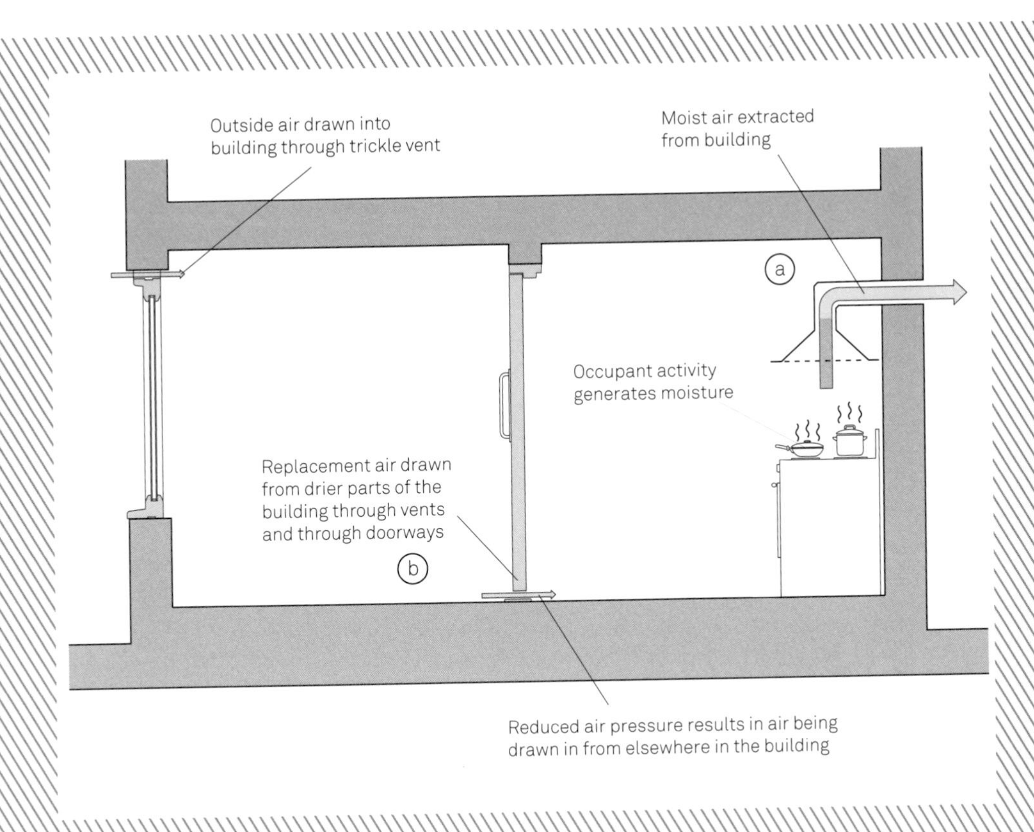

FIGURE 3-07 CONTROLLING MOISTURE LEVELS BY EXTRACTION AT SOURCE

It is also important to consider the thermal mass of the element in relation to the heating pattern. A structure with high thermal mass (e.g. a concrete slab roof) in a building heated intermittently (say, a church) may be at risk of surface condensation, because its surface will heat up slowly compared with the air around it, resulting in air being cooled at the surface.

Buildings with higher humidity levels (see: Human activity p. 80) will need surface temperatures higher than those with lower internal humidity, and will consequently require more energy for heating or – preferably – better insulation levels.

TEMPERATURE FACTOR

The extent of cooling at a surface, usually that of a junction or thermal bridge, is expressed by the temperature factor f, which is the ratio of the difference between the surface temperature and the outside air temperature, to that of the inside and external air temperatures.

$$f = \frac{T_{si} - T_e}{T_i - T_e}$$

where:

T_{si} is the internal surface temperature

T_i is internal air temperature

T_e is the external air temperature

The three temperature values required to calculate the temperature factor are obtained by numerical modelling of the junction (see: chapter 1; Heat transfer at junctions p. 16). For a well-constructed junction with little heat loss the internal surface temperature will be very close to the internal air temperature, resulting in a temperature factor close to 1.0. Constructions with serious thermal bridging problems will have temperature factors approaching 0.5. For UK conditions a temperature factor of 0.75 or greater is sufficient to prevent surface condensation and mould growth.

The temperature factor can be used to calculate the internal surface temperature for known internal and external air temperatures. If the internal relative humidity is known then it is possible to calculate the surface relative humidity and check the risk of surface condensation.

For further details of temperature factors at junctions see BRE Information Paper 1/06.

Controlling interstitial condensation

Interstitial condensation is a hidden but damaging moisture problem which occurs within the building fabric and can cause substantial damage, often without any visible signs, for example:

* Corrosion of metal components – such as fixings or framing – which can lead to structural failure
* Damage to electrical installations
* Rot and decay of timber and timber-based products, leading to loss of structural strength and buckling of timber-based panels

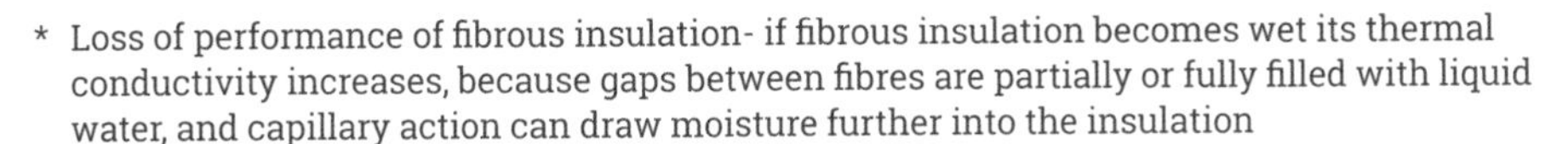

* Loss of performance of fibrous insulation- if fibrous insulation becomes wet its thermal conductivity increases, because gaps between fibres are partially or fully filled with liquid water, and capillary action can draw moisture further into the insulation
* Damage to roofs - condensation forming beneath the waterproofing membrane can result in blistering
* Mould growth within the construction – for example, in belongings stored in voids in the roof space in the conventional UK cold pitched roof

How and where interstitial condensation occurs

To understand how interstitial condensation occurs, and how to avoid it, we must examine the behaviour of water vapour in solids.

Water vapour and solids

Water vapour moves through the building fabric by diffusion (see above: How water vapour moves p. 78). Wherever there is a difference in vapour pressure between the inside and outside of a building – which is the case for virtually all conditioned buildings – there will be a vapour drive across the building element, with water vapour moving from the higher pressure area to the lower pressure area. The rate at which vapour moves through the element is affected by the size of the pressure difference, and the vapour resistance of materials within the element.

Where all the materials in a building element have similar vapour resistances water vapour will enter and leave the element at roughly the same rate. But where one of the materials has a higher vapour resistance (such as plywood sheathing within a timber-framed wall), the vapour will build up behind it, resulting in a higher vapour pressure at that interface.

As the inside and outside surfaces will be at different temperatures there will be a temperature gradient across the element, often with large falls across layers of thermal insulation. The SVP will change with the temperature, decreasing from one side of the element to the other, and also dropping significantly across insulation layers.

With some combinations of materials, vapour pressures and temperature, a low SVP can coincide with a build-up of vapour at a layer with a high vapour resistance, resulting in the SVP being reached and the remainder of the water condensing within the construction.

We can see this process at work if we look at a basic timber-framed wall, which has an internal lining, fibrous insulation between timber studs, plywood sheathing (to provide racking strength) and a weather-proof cladding. It is easier to look at the relationships in terms of the dew point temperature, rather than SVP.

Figure 3–08 shows the temperature gradient and the dew point temperature gradient across the element – based in this case on winter conditions (heating season) in the UK. On the internal face the actual temperature and the dew point temperature are separated by 5°C, but both drop substantially across the insulation layer, with the actual temperature hitting the dew point temperature at the interface of the insulation and the plywood, predicting the occurrence of condensation at the inside face of the plywood.

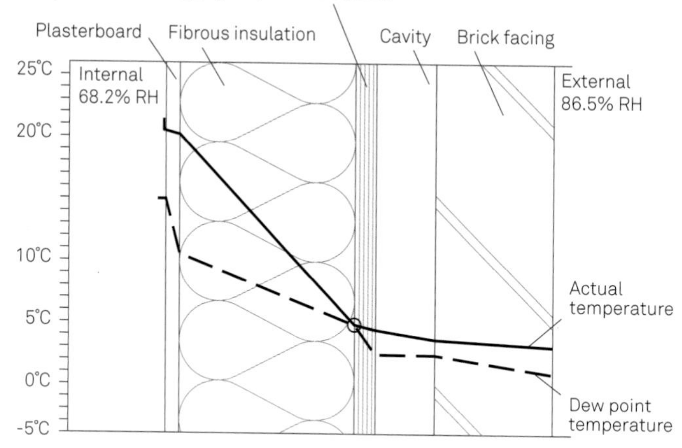

FIGURE 3-08
CONDENSATION RISK IN A TIMBER FRAMED WALL

Why does condensate occur at that interface? The interface is a classic example of the condensation position: the plywood is on the cold side of the insulation, and will not be much warmer than outside air, which means the SVP will be low. At the same time, the high vapour resistance of the plywood (hundreds of times greater than that of the adjoining insulation) results in a high vapour pressure at the interface. The combination of high vapour resistance on the cold side of the insulation gives interstitial condensation.

The vapour drive across a building element will change seasonally as environmental conditions change. In some cases, that will result in the evaporation of some or all of the condensate which has formed within the element. Any assessment of the potential damage from condensation must take account of any year-on-year build-up, but also recurring condensation which evaporates over the course of the year.

Bulk air movement

Bulk air movement through gaps and cracks in the fabric can produce more vapour movement than diffusion, leading to high vapour pressures within building elements. This is particularly a problem when air from humid spaces (such as bathrooms or shower rooms) can reach cold voids, such as lofts with insulation fitted between and across ceiling joists. The resulting combination of high moisture levels and low temperatures makes condensation highly likely.

Assessing the risk of interstitial condensation

The likelihood of interstitial condensation forming within a given construction should be assessed: this is essential for buildings that are expected to have high amounts of water vapour generation (e.g. laundries and dwellings). There are two main methods of assessing interstitial condensation risk: **steady state** and **transient**. For both methods the basic procedure is the same: construct a computer model of the element, define the environmental conditions and run the analysis, then use the results to determine whether the proposed construction is likely to perform acceptably.

Steady state analysis

A steady state analysis – sometimes referred to as Glaser method[4] – proceeds by calculating the vapour pressure and SVP at each interface in the element (i.e. a plane on which two materials abut, such as where fibrous insulation between studs rests against plywood sheathing), based on the vapour resistance and thermal resistances of the materials, and the internal and external temperature and vapour pressure. Condensation is predicted to occur at any interface where the vapour pressure is the same as the SVP. It is then possible to calculate the amount of condensate likely to be deposited for a given period: this is expressed in grams per square metre (g/m^2).

The analysis is conducted for a year, using mean monthly figures for internal and external temperature and relative humidity. (Using relative humidity makes it easier to find local weather station data to use in analysis). The software reports how much – if any – condensate is predicted at each interface for each month, and gives the cumulative, annual figures, allowing for the effect of evaporation.

For any construction there are three possible outcomes:

* No condensation is predicted in the construction
* Condensation is predicted during some months, but it evaporates at other points during the year, so there is no year-on-year build-up of condensation
* Condensation is predicted, with a year-on-year build-up

The first condition constitutes a 'pass' and the third a 'fail', but the second condition requires careful consideration to determine whether the predicted condensation constitutes a hazard. The key results to consider are: how much condensate is predicted to be deposited; how long would it remain; and which materials would be in contact with the liquid water. For example, 5 g/m^2 on the face of brickwork for a month is likely to be acceptable, but 200 g/m^2 on the face of a timber-faced board for several months would be unacceptable (leaving aside the fact that this amount of condensate would run and drip).

Steady state analysis is straightforward and quick, and useful for identifying potential condensation problems at the design stage. It can also be used to compare the performance of different constructions to determine which would be least susceptible to condensation. However, like any model, it has limitations, the first being that it ignores some of the more complex but significant behaviours of water, notably:

* Capillary action
* The taking up of liquid water into the body of a material (absorption) or its surface (adsorption)
* Phase change as condensation and evaporation take place
* The effect of hygroscopic materials

The second limitation is that the method is unable to model short-term, transient changes, particularly where there are large diurnal variations in temperature and humidity.

Transient analysis

To overcome the limitations of steady state analysis more complex methods of analysis have been developed, which: (i) take account of the phenomena steady state analysis ignores; and (ii) look at results over a much shorter period.

Much of the underlying methodology is similar, but the analysis is carried out on an hourly basis, calculating the moisture balance for one hour, then using those results as the starting point for the next hour. It also considers the interaction between moisture movement and heat, the effect of liquid water transport by capillary flow and the effects of solar radiation, precipitation and wind speed on the construction.

The results of transient analysis are more accurate, particularly for thermally massive constructions and projects with significant diurnal variation environmental conditions. Transient analysis is particularly useful for assessing remedial action on existing building elements.

Despite its complexity, transient analysis still does not address the movement of water vapour resulting from air movement within building elements.

Gathering data

One challenge for condensation risk analysis is obtaining suitable climatic and material data. For steady state analysis it is usual to employ temperature and relative humidity data from a weather station near to the site: most software comes with a number of suitable data sets. In the UK these are almost all airports or RAF stations: those being the only places where the necessary data is collected.

The hourly data required for transient analysis may be more difficult to find: the more complex the analysis the more important it is to find representative climate data. Again, some is provided with software, and professional bodies such as the Chartered Institute of Building Services Engineers (CIBSE) have also developed representative climate data.

A further complexity is that observational data is innately historic, reflecting the past climate, but not the future climate: CIBSE and the UK Meteorological office have therefore collaborated to develop predicted weather data for analysing building performance under probable future conditions.[5]

Preventing interstitial condensation: climate conditions

To prevent harmful interstitial condensation occurring within a building element we need to address the three components which interact to cause it, namely: the external climate, the conditions within the building and the building fabric.

External climate

The external climate is a given: there is very little we can do to change it. (That is not to say human activity does not affect the climate, but when considering interstitial condensation risk the climate is what it is, or at least, what it is predicted to be.) Positioning a building to take advantage of topography and orientation – perhaps to make the best use of solar gain, or to reduce the effect of prevailing winds – will improve its overall performance, but will not usually make a substantial difference to the risk of interstitial condensation.

However, there are two conditions which should be considered:

* Solar gain – direct sunlight can raise the external surface temperature of walls and roofs above the air temperature. This can drive moisture towards the building interior where it may condense against a layer with a high vapour resistance. Transient condensation risk analysis (see above: Transient analysis p. 91) should be undertaken to determine whether that will be a problem and, if it is, to evaluate the effect of design changes.

* Night radiative cooling – clear night skies can produce external surface temperatures on roofs which are much lower than the air temperature (see chapter 1: Assessing heat transfer through the building fabric p. 13), resulting in interstitial condensation on the underside of the roof covering. Absorptive fleeces may use used on the underside of uninsulated metal sheet roofs to prevent condensate running or dripping.

Internal conditions

The two key parameters are humidity and temperature. High moisture levels must be avoided: the principles set out above (see: Controlling moisture levels p. 86) are equally valid for preventing interstitial condensation. Moisture should always be extracted as close as possible to the point of generation to ensure that moisture levels cannot build up in unheated or colder parts of a building.

The heating regime needs to be matched to the thermal mass of the fabric: thermally massive fabric should not be coupled with intermittent heating. Where the building interior is cooled or fully conditioned for comfort there is a possibility that an outside-to-inside vapour drive may be created, which could result in interstitial condensation (see box below: Moisture damage in manufactured homes).

MOISTURE DAMAGE IN MANUFACTURED HOMES

Towards the end of the 20th century, many owners of 'manufactured homes' (UK: mobile homes) in the south-east of the USA were experiencing moisture-related problems of decay, with wall and ceiling panels buckling and rotting, and floors failing. A research project identified a number of issues which were combining to cause the problems.

Hot, humid climates such as those in Florida and Southern Georgia have high dew point temperatures. Building occupiers often run large air-conditioning units to cool the interior; however, oversized units cool before adequate dehumidification has occurred, resulting in surface condensation.

At the same time, where there is air leakage and diffusion from outside, the vapour is transported towards the building interior. As the internal linings are often finished with vinyl – which has a high vapour resistance – condensation builds up behind the finishes, damaging both the finishes and the structure. The problem is exacerbated by oversized air-conditioning units which produce negative pressure gradients, drawing moist external air into dwellings.

With the causes identified, manufacturers were able to tackle many of the issues around air leakage, reducing the occurrence of problems.

Preventing condensation: building fabric

The principle way to address interstitial condensation via the building fabric itself is to prevent a build-up of water vapour at any interface on the cold side of any thermal insulation (see box below: Warm side, cold side), by selecting and arranging materials with regard to their hygroscopic properties, particularly their thermal resistances and vapour resistances.

WARM SIDE, COLD SIDE

When discussing interstitial condensation risk it is common to refer to the warm side and cold side of an element, because the vapour drive is usually from warm to cold. Which side of an element qualifies as the cold side depends mainly upon the climate and the conditioning of the building. For the UK and most of Northern Europe the primary vapour is from inside to outside of the insulation during the winter: there may be a slight inward vapour drive during the summer in some conditioned buildings.

For the USA and Australia, those regions with colder winters have an outward vapour drive to consider, but in warmer areas (the southern states of the USA, and the North of Australia) it is the vapour drive inward which needs to be considered. And there are some locations, sitting on the boundary, where there is no fundamental drive to consider, but a balance of in and out, depending on the season.

The ideal arrangement is a construction in which the vapour resistance of each material decreases in the same direction as the vapour drive, which will usually be from the warm side of the construction to the cold side. Ensuring each layer has a lower vapour resistance than the one before it allows water vapour to diffuse out of the construction as fast as it diffuses in (see Figure 3–07).

Practically, where the vapour resistances of materials within an element are low, there is little risk of interstitial condensation, even if the high–low gradient is not observed at every interface, because each material presents very little resistance to vapour transfer (e.g. a timber-framed wall with vapour permeable wood fibreboard as sheathing).

Many constructions contain materials with high vapour resistances (e.g. plywood, sheet metal, and bituminous and plastic sheets). Although those constructions fall short of the ideal they are not at risk of condensation provided the materials with high vapour resistance are to the warm side of the thermal insulation.

The timber-framed wall in Figure 3–08 has a condensation risk because the plywood sheathing – which has a high vapour resistance – is on the cold side of the thermal insulation. It is the combination of high vapour resistance and low temperature which creates the risk. As a rule of thumb, we can say that a layer with a high vapour resistance on the cold side of thermal insulation should be considered to be at risk of interstitial condensation. (It is almost true to say that without the introduction of thermal insulation there would be no interstitial condensation risk.)

The simplest solution is to move the material to a different place in the construction (e.g. putting the plywood sheathing on the inside of the framing – an unconventional but feasible change), or replace it with one which has similar structural properties but lower vapour resistance (e.g. replacing plywood with oriented strandboard [OSB].

However, there are many constructions where neither relocation nor substitution is feasible. For those constructions we can adopt one of two solutions:

* Changing the balance of vapour resistances within the construction by using a VCL
* Removing moisture from the construction by means of ventilation

Vapour control

A VCL is formed by a coating or membrane with a high vapour resistance, such as polyethylene sheet or a composite membrane with an encapsulated aluminium foil. The VCL is installed on the warm side of the insulation, where it slows the rate of water vapour transfer through the construction. (Note that, because VCLs rarely prevent all vapour transfer, they are no longer referred to as 'vapour barriers'.) For some constructions, such as metal skinned insulated panels, there is no need to introduce a separate VCL: simply sealing the panel junctions will provide the necessary vapour control. The effect of a VCL on vapour conditions can be seen in Figure 3–09, which shows conditions through a timber-framed wall similar to that in Figure 3–08:

* Without a VCL the predicted temperature reaches the dew point at the interface between the insulation and the plywood, resulting in condensation
* With a VCL, the dew point at that interface is lower, and the predicted temperature does not reach the dew point

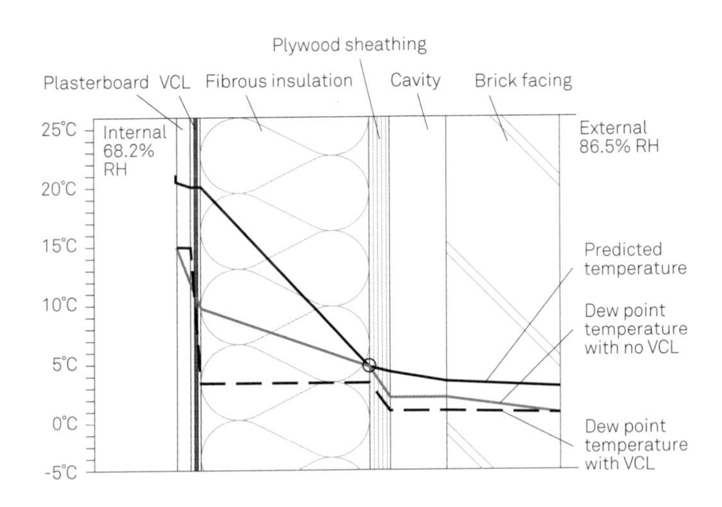

FIGURE 3-09
PREVENTING INTERSTITIAL CONDENSATION WITH A VAPOUR CONTROL LAYER

The vapour resistance required of a VCL will depend on the environmental conditions and the rest of the construction and can be determined by condensation risk analysis.

Good detailing and accurate installation (laps, junctions and service penetrations must all be properly formed and sealed) are essential to the formation of an effective VCL, but the best results are achieved by a systematic approach to vapour control for the building as a whole. One method is to identify a continuous plane for vapour control within all elements of a building, giving particular attention to junctions between elements. VCLs have the additional benefit of preventing air leakage into the construction, which will prevent vapour transmission by mass transfer and so reduce condensation risk.

Ventilation for the fabric

There are some constructions where it is not practicable to provide a VCL (e.g. renovations where existing finishes are being retained). This can mean there is a risk of condensation if any of the layers to the cold side of the construction have high vapour resistance. In the absence of a VCL the only solution is to ventilate the construction behind the layer with high resistance and so prevent the build-up of water vapour. Figure 3–10 shows this approach for two constructions.

Figure 3–10a shows the refurbishment of a pitched roof, where thermal insulation is added on the slope, but the bituminous underlay is retained; water vapour can be removed from beneath the underlay by forming vented cavities between the insulation and underlay. In Figure 3–10b, there is the risk of condensation on the underside of a sheet metal roof covering; the use of a vapour open membrane with a plastic spacer mesh forms an airspace which vents water vapour to the atmosphere and allows condensate to drain away.

The introduction of an air current can adversely affect the thermal performance of the construction (see chapter 2: Air in the building fabric p. 52) but that is the price of preventing damage from condensation. Of course, a better solution would be to employ a different construction, which does not require an air current to remove the moisture.

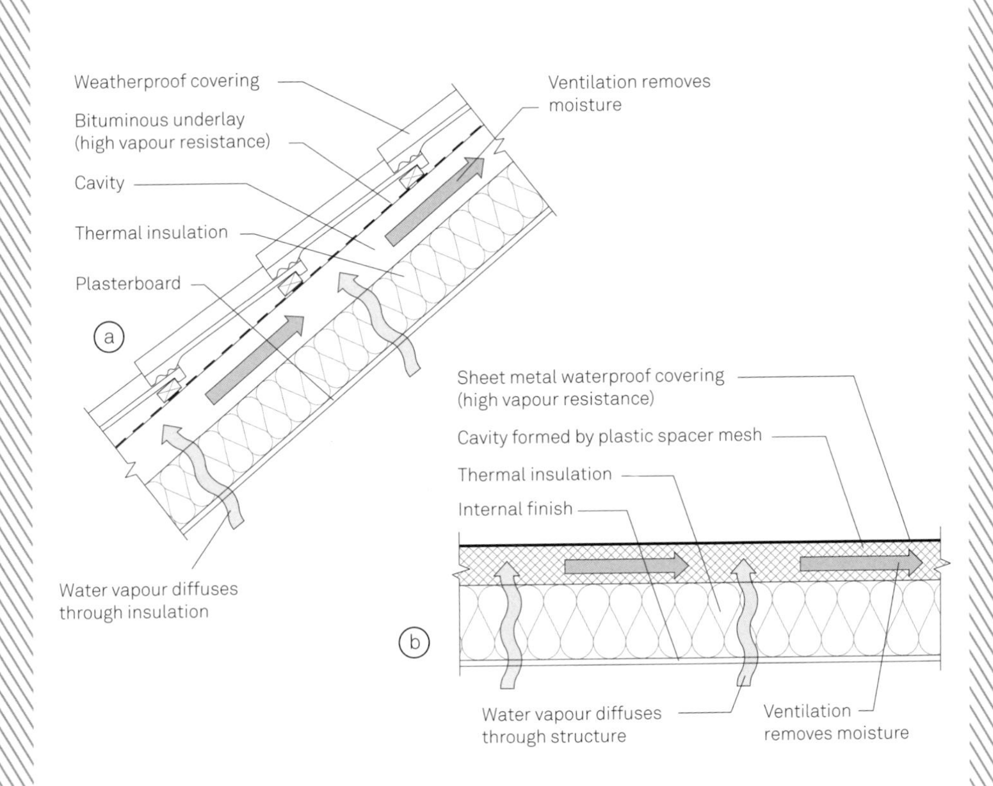

FIGURE 3-10
VENTILATION TO PREVENT INTERSTITIAL CONDENSATION WHERE BUILDING ELEMENTS CONTAIN LAYERS WITH VERY HIGH VAPOUR RESISTANCE

Moisture in the bigger picture

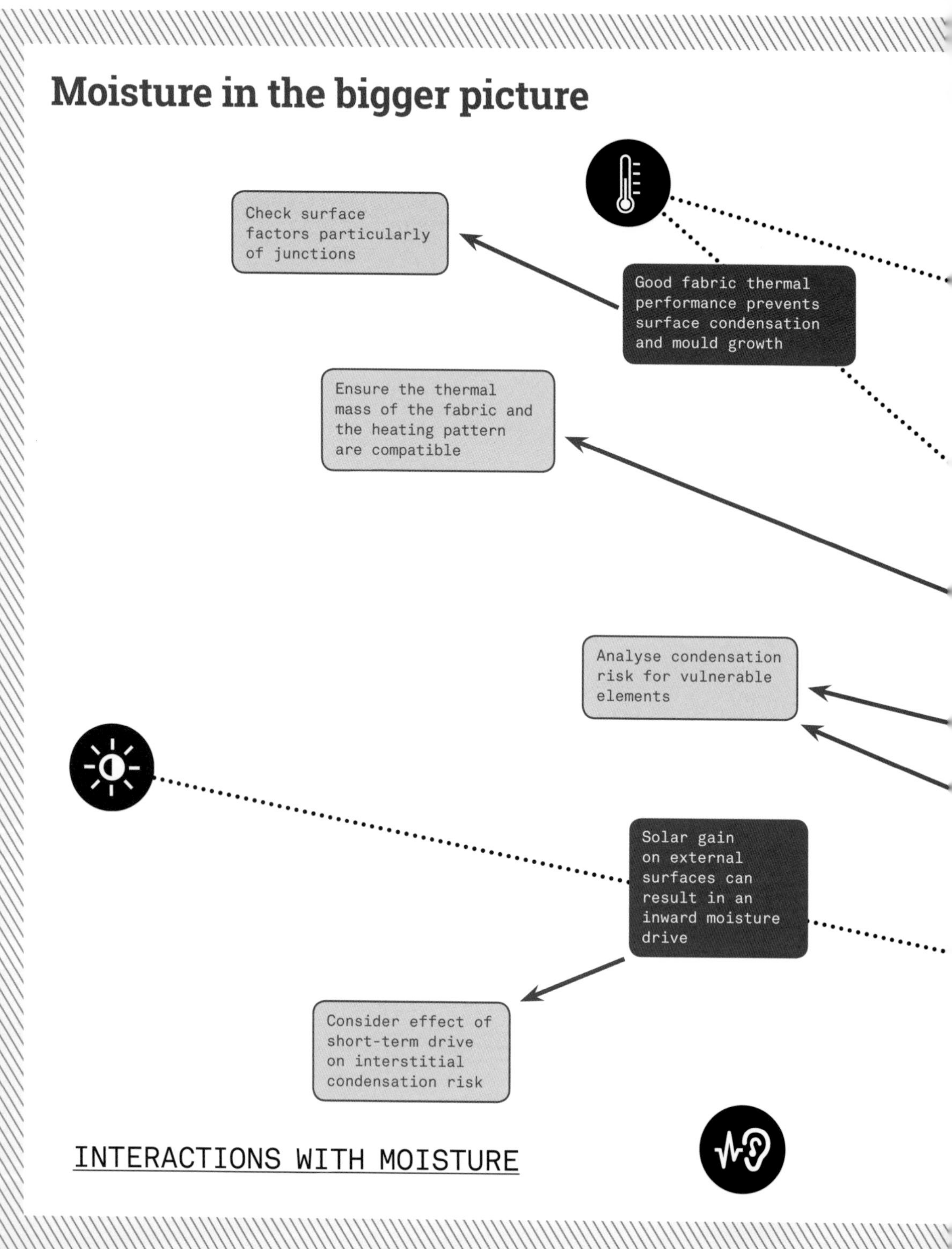

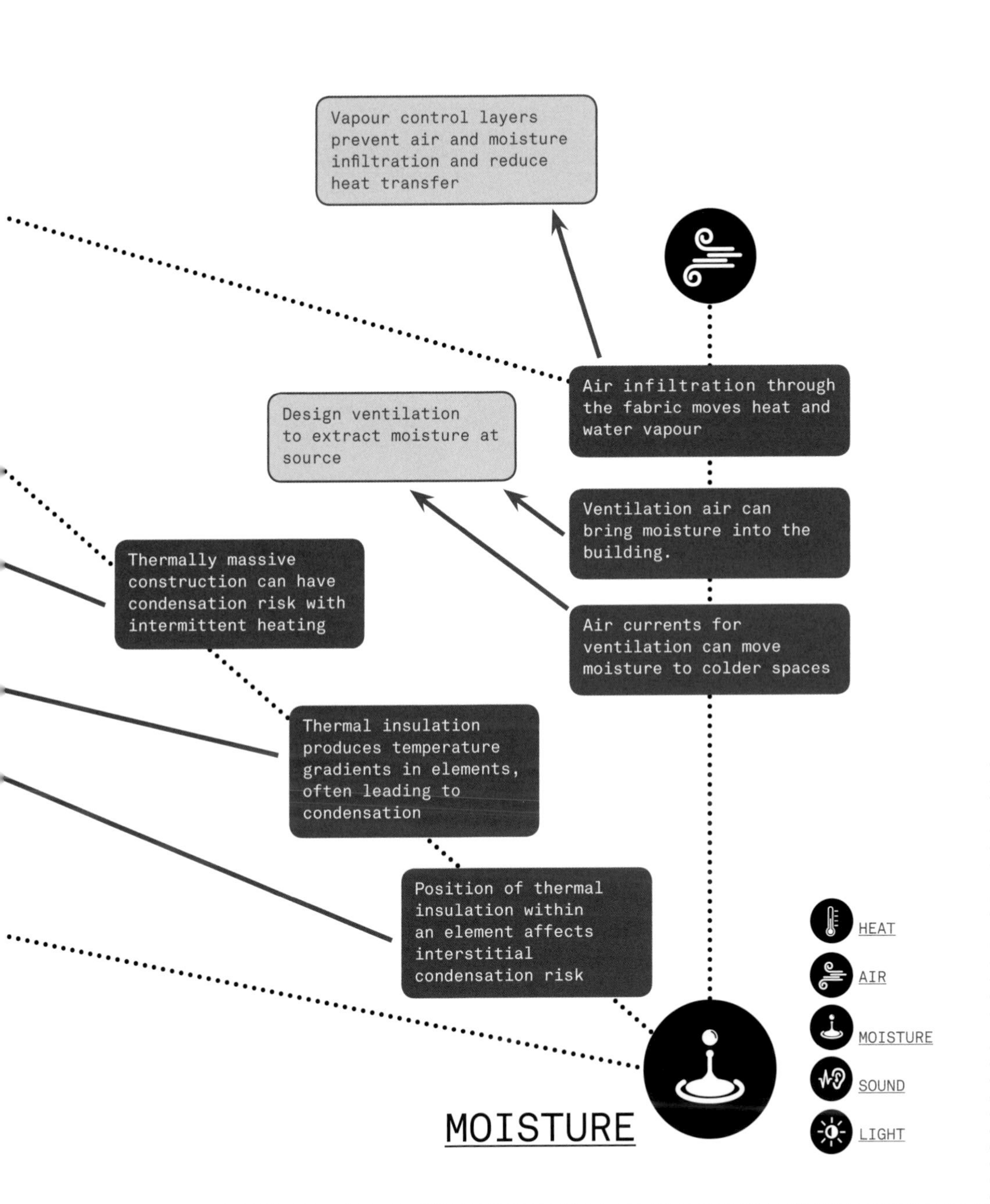
Vapour control layers prevent air and moisture infiltration and reduce heat transfer
Design ventilation to extract moisture at source
Air infiltration through the fabric moves heat and water vapour
Ventilation air can bring moisture into the building.
Air currents for ventilation can move moisture to colder spaces
Thermally massive construction can have condensation risk with intermittent heating
Thermal insulation produces temperature gradients in elements, often leading to condensation
Position of thermal insulation within an element affects interstitial condensation risk
MOISTURE
HEAT
AIR
MOISTURE
SOUND
LIGHT

04

Sound

DEFINITION:
SOUND: PRESSURE WAVES PROPAGATED THROUGH FLUIDS AND SOLIDS, THAT ARE DETECTED BY THE ORGANS OF HEARING.

Air vibrates. It is continually set in motion by natural forces (the wind, the sea), by animals and by people. The human ear can detect those vibrations and the brain understands them as sound. Sound facilitates communication and is a medium for pleasure and joy. Yet sound can also create a nuisance and be detrimental to health and wellbeing.

When designing buildings it is not possible to eliminate noise, but by attending to the physics of sound we should be able to strike out (or at very least, reduce) the nuisance, make clear the communication and so enlarge the joy.

There are two main issues to address.

Sound from outside, or from another part of a building, travels into a space, disturbing the occupants. This intrusive sound may be caused by aircraft or traffic noise, footfalls on bare floors, televisions in adjoining dwellings, or the operation of building services. Intrusive sound can affect health and quality of life. The acoustic characteristics of the surfaces bounding a space can result in reverberation, which reduces the intelligibility of speech and makes communication more difficult. Excessive reverberation is often a problem in schools and workplaces.

With those issues in mind, this chapter will introduce the basic principles of sound and human hearing, consider the common problems caused by sound, and then outline measures that can be taken to reduce them.

Note, however, that this chapter can only address the common applications that apply to the majority of buildings; it does not attempt to deal with the special conditions required for concert halls, recording studios and the like, which require the knowledge and skills of acousticians, although, in virtually all cases, the techniques they employ are all based on the fundamental principles described here.

The fundamentals of sound

The physical phenomenon which we perceive as sound consists of waves of alternating high and low pressure travelling through air or another medium. The speed at which the pressure waves travel through a medium depends on the density of the medium (higher density reduces speed) and its stiffness (greater stiffness increases speed). Table 4.1 shows the speed of sound in common materials.

TABLE 4.1 SPEED OF SOUND IN COMMON MATERIALS

MATERIAL	SPEED m/s
Air (sea level, 0°C)	343
Brick	4176
Glass	5640
Hardwood	3962
Softwood	3300–3600
Stainless steel	5790
Water	1433

SOURCE: DATA FROM WWW.ENGINEERINGTOOLBOX.COM.

The two main characteristics of a sound wave are:

* Frequency – the number of cycles of high and low pressure per second. Frequency is expressed in hertz (Hz). In Figure 4–01 the solid line shows a wave pattern which has one oscillation in a second, so has a frequency of 1 Hz. The dotted line shows a wave with two oscillations in a second, so has a frequency of 2 Hz. A sound wave with one hundred cycles of high and low pressure in a second has a frequency of 100 Hz.
* Sound pressure – the increase in air pressure produced by the wave, which is measured in pascals (Pa) and is associated with the loudness of the sound. Both waves in Figure 4–01 have the same pressure range.

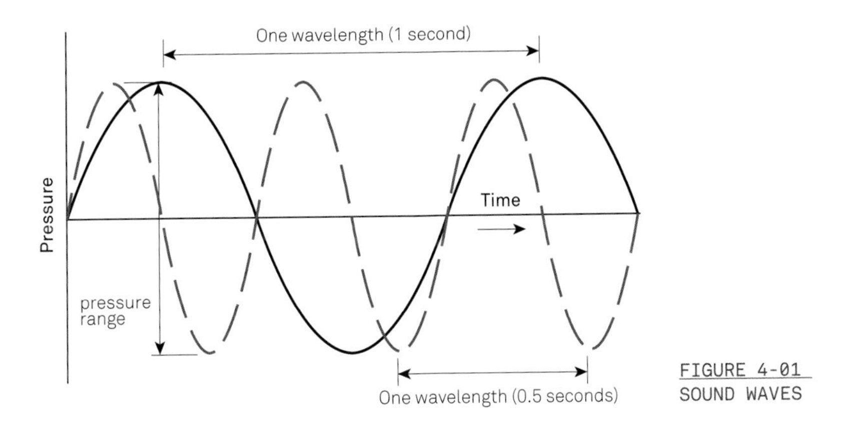

FIGURE 4-01
SOUND WAVES

Human hearing

The human ear (Figure 4–02) is a complex physical mechanism which converts pressure waves moving through air into electrical impulses that are passed to the brain to be analysed and perceived as sound.

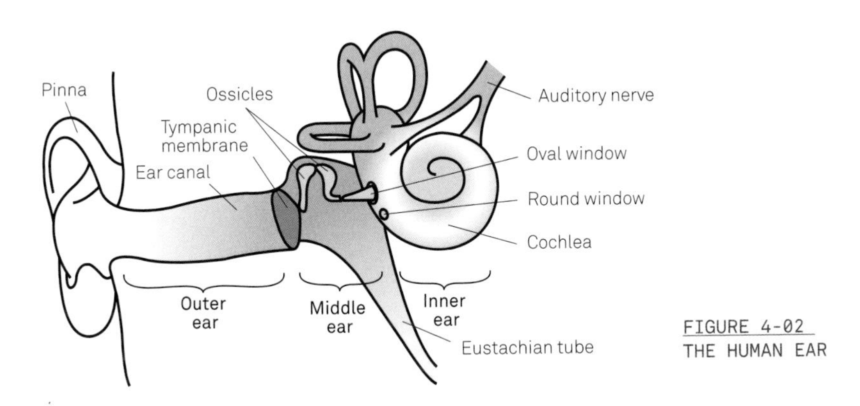

FIGURE 4-02
THE HUMAN EAR

The outer ear consists of the ear canal, which extends from the external flap of cartilage and skin we commonly refer to as the ear, to the ear drum or tympanic membrane. The middle ear is an air-filled cavity containing three small bones (the ossicles) that link the tympanic membrane to the cochlea, which is part of the inner ear. The cochlea is filled with fluid and lined with thousands of fine hair cells, which are connected to the auditory nerve.

Pressure waves in the air travel down the ear canal and set the tympanic membrane vibrating. The vibrations are passed through the ossicles (which help to concentrate the energy of the sound waves) to the fluid in the cochlea, which sets the hair cells vibrating. The cells respond to the vibration by sending electrical impulses through the auditory nerve to the brain. Hair cells in different parts of the cochlea respond to different frequencies of vibration, giving sensitivity to pitch.

Humans can typically hear frequencies of 20-20,000 Hz, although the range differs between individuals, and there is usually loss of sensitivity at the upper end of that range with age, as hair cells fail.

The mechanisms of sound transmission

There are two mechanisms by which sound travels into and around buildings:

* **Airborne transmission** – Sound can travel into a space in a building directly through the air, by passing through gaps and spaces in the structure, including windows, ventilation openings and ducts (see Figure 4–03). It can also travel indirectly, as sound waves striking one side of a building element set it in motion, producing vibrations in air on the other side.
* **Impact transmission** – When an object strikes the building fabric the energy of impact sets up vibrations within the fabric, which travel through it and set air in adjacent spaces vibrating, creating sound waves. A common source of impact sound is footfall on floors, but it is also produced by shutting doors (as they bang against the frame), the operation of electrical switches and sockets, and, of course, rainfall on roofs.[1]

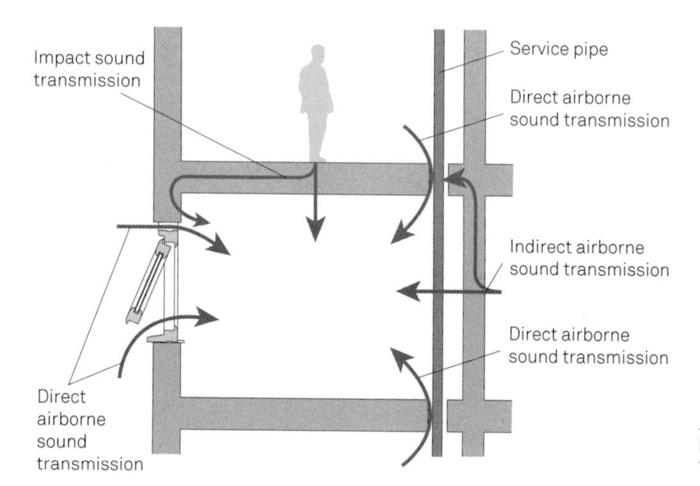

FIGURE 4-03
MECHANISMS OF SOUND TRANSMISSION

When sound travels through the building fabric it often takes more than one path: impact sound generated on a floor may also set up vibrations in the supporting walls; vibrations in wall cavities can travel round the ends of walls into adjoining cavities, while pipes and ducting which pass through elements also provide routes for sound transmission. Addressing this **flanking transmission** of sound is a crucial part of tackling sound problems.

The surfaces of a space can reflect sound waves, prolonging the sound beyond its normal duration. This is shown in Figure 4–04, where the reflected sound reaches the listener slightly after the sound coming directly from the source.

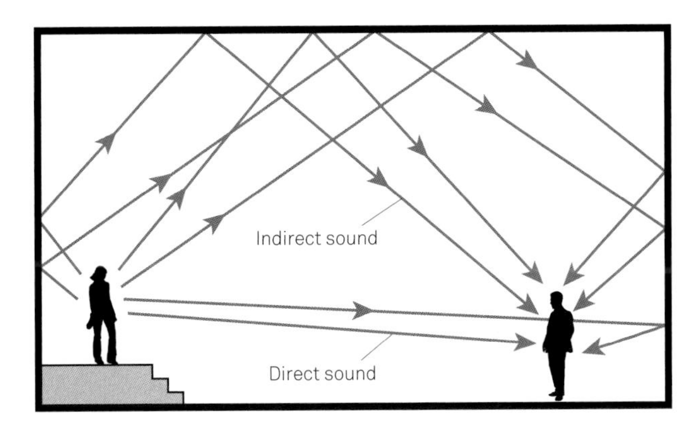

FIGURE 4-04
REVERBERATION AND INTELLIGIBILITY OF SPEECH

Measuring sound

Loudness

The human ear is sensitive to a wide range of sound pressures, from a just audible sound at 20 micropascals (20 µPa) up to the threshold of pain at about 200 Pa. The pressures at the top of the range is about ten million times greater than those at the lower end.

The sound pressure level (SPL) is used to represent this range of sound pressures in a workable manner. The SPL is a ratio of the sound pressure being measured compared with the reference sound pressure of 20 µPa. Because the scale is logarithmic (that it, expressing the ratios in powers of ten) a sound which is 10 decibels (10 dB) greater than another will be twice as loud. Figure 4–05 shows the SPL for a range of common sound sources.

The SPL experienced by someone hearing a sound is affected by the distance from the source. There are two reasons for this:

* Some of the energy of the sound wave will be absorbed by the medium through which it is travelling.
* As a sound wave travels out from the source it expands across an ever-increasing face, in the same way as ripples from a thrown stone spread across the surface of a pond and die away. The further the sound wave is from the source, the larger its surface area, and the less energy there is at any point (the principle is the same as shown in Figure 5–04 for light). Each doubling of distance from the source decreases the SPL by 6 dB.

The perceived loudness of a sound is affected by its frequency, because the ear is more sensitive to sounds in the mid-frequencies – approximately 1000–4000 Hz – than to low and high frequency sounds. So, for example, a 1000 Hz sound with a SPL of 70 dB will seem as loud as a 100 Hz sound at 75 dB. To allow for that variation in sensitivity, measured sound levels are weighted to take more account of middle-range frequencies than low and high frequencies. The commonest weighting method is the A weighting curve, which reduces the contribution to loudness of frequencies below 1000 Hz. In a noise assessment a value quoted as a 'dBA' value, (or dB(A)) rather than just dB, has been adjusted using the A weighting curve.

The acoustic performance of buildings is also different across frequency ranges. In order to give values which convey the performance across the whole frequency range, SPLs are measured in frequency bands, using either octave bands (where the frequency at the top of the band is twice that at the bottom of the band), or one-third octave bands (Figure 4–06). The results across the bands are then combined into one value, which is usually weighted to emphasise the significant frequency bands for an application: for example the C_{tr} correction to airborne sound transmission results is used to emphasis the effect of lower frequencies in traffic noise.

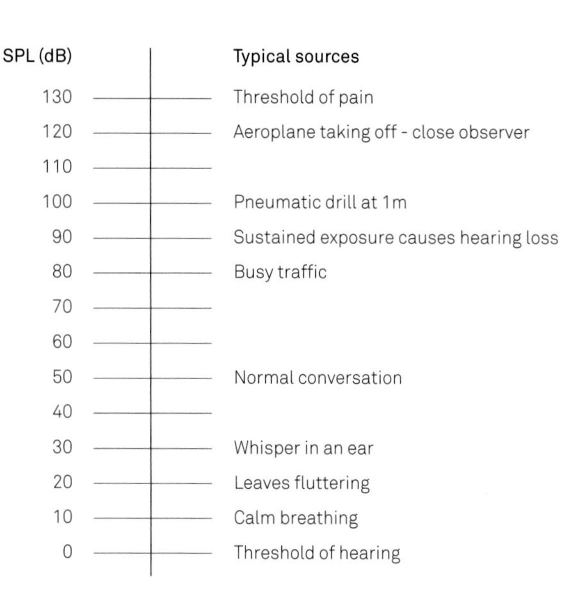

FIGURE 4-05 SOUND PRESSURE LEVELS FOR COMMON SOURCES OF SOUND

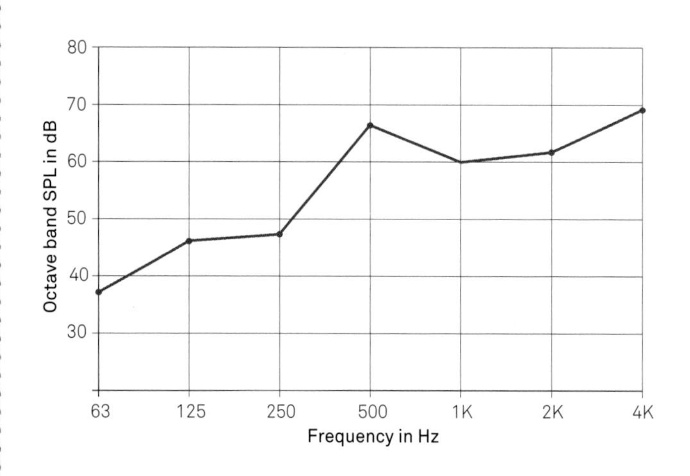

FIGURE 4-06 A TYPICAL FREQUENCY GRAPH SHOWING OCTAVE FREQUENCY BANDS

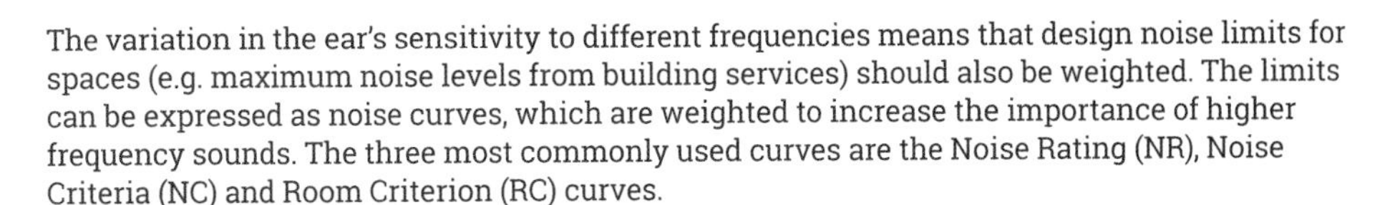

The variation in the ear's sensitivity to different frequencies means that design noise limits for spaces (e.g. maximum noise levels from building services) should also be weighted. The limits can be expressed as noise curves, which are weighted to increase the importance of higher frequency sounds. The three most commonly used curves are the Noise Rating (NR), Noise Criteria (NC) and Room Criterion (RC) curves.

Sound transmission

The airborne sound performance of an element is measured as the difference between the loudness of a sound generated on one side of an element and its loudness measured on the other side. Elements which transmit less sound produce higher differences; those which transmit more sound produce lower differences.

Airborne sound tests are carried out using a standard noise source in the source room, while microphones in the receiving room measure the loudness of the noise. The sound transmission of different frequencies varies, so measurements are taken across multiple frequency bands and compared with a standard curve to obtain an average performance figure. The key measures are:

* R_w – or the weighted sound reduction index – a laboratory measure of airborne sound insulation based on a standard test suite and measured in dB
* D_{nTw} – or the weighted standardised level difference – a site measure of airborne sound insulation, adjusted for reverberation, measured in dB
* C_{tr} – a correction factor which increases the contribution of low frequency sound to the D_{nTw} value

Airborne sound measurements record sound reduction, so higher values for R_w and D_{nTw} are better.

Impact sound performance is expressed as the loudness of a sound generated by a standard impact on the building structure. Elements which generate less sound have lower sound levels, while those which generate more sound have higher sound levels.

Impact sound tests are carried out with a tapping machine, which strikes the floor surface with a known force, while microphones in the receiving room measure the sound level. Measurements are taken across multiple frequency bands and compared with a standard results curve to obtain an average performance figure. The key measures are:

* L_{nW}– the weighted normalised impact sound pressure level – a laboratory measure (in dB) of impact sound based on a standard test suite
* L_{nTw}– the weighted standardised impact sound pressure level – a site measure of impact sound performance (in dB), adjusted for reverberation time

As both L_{nW} and L_{nTw} measure the level of sound transmission, lower values are better

The change in impact sound level produced by a surface treatment is expressed as ΔL_w and is measured in decibels: higher values indicate greater improvement.

In situ testing of a completed building element remains the most accurate method of determining its performance; however, if the element fails to achieve the specified performance level then disruptive remedial works will usually be required. To avoid such problems it is common to rely on systems which laboratory testing shows can achieve well above the required standard.

There are predictive calculations, but these are not currently widely used. Like any other calculation or modelling method, the final result depends on the standard of construction. And with sound transmission, small gaps in the fabric can be detrimental to performance.

Reverberation and absorption

Reverberation is the length of time it takes for a sound to die away, once the source has stopped. The standard measure of reverberation time, (R_{T60}) is the time taken for a sound to decay by 60 dB, measured in seconds. (Testing for 60 dB decay is not practicable in the field, so the first 20 dB decay is measured and the 60 dB time extrapolated. The results should be labelled as $R_{T60(T20)}$.)

Reverberation is affected by the amount of sound absorbed by the surfaces of a space. The proportion of sound falling on a surface which is absorbed by the surface is measured by the absorption coefficient, α:

* 1 represents a surface which absorbs all sound falling on it
* 0 represents a surface which reflects all sound falling on it

A fully open window has an absorption coefficient of 1, because it will not reflect any of the sound which reaches it. Table 4.2 lists absorption coefficients for common materials. Note the coefficients vary with the frequency of the sound.

TABLE 4.2 ABSORPTION COEFFICIENTS FOR COMMON SURFACES

CONSTRUCTION	SOUND ABSORPTION COEFFICIENTS, α IN OCTAVE BANDS (Hz)				
	250	500	1000	2000	4000
Fair-faced concrete or plastered masonry	0.01	0.01	0.02	0.02	0.03
Fair-faced brick	0.02	0.03	0.04	0.05	0.07
Painted concrete block	0.05	0.06	0.07	0.09	0.08
Windows, glass façade	0.08	0.05	0.04	0.03	0.02
Glazed tile/marble	0.01	0.01	0.01	0.02	0.02
Hard floor coverings (e.g. linoleum, parquet) on concrete floor	0.03	0.04	0.05	0.05	0.06
Soft floor coverings (e.g. carpet) on concrete floor	0.03	0.06	0.15	0.30	0.40
Suspended plaster or plasterboard ceiling (with large air space behind)	0.105	0.10	0.05	0.05	0.05

SOURCE: BUILDING REGULATIONS APPROVED DOCUMENT E 2004, TABLE 7.1.

The total sound absorption of a room is calculated by multiplying the area of each surface by its absorption coefficient, α. Its reverberation time depends on its volume (V) and its total sound absorption (A) and can be calculated using the formula:

$$R_{T60} = 0.16\ V/A_T$$

The absorption coefficients for materials vary with the frequency of the sound, so the reverberation time calculations must be undertaken for several different frequency bands, typically 250, 500, 1000, 2000 and 4000 Hz.

Sound, people and buildings

Sound can have a significant impact on the occupants of buildings, affecting their health, their ability to work and their quality of life.

Sound and health

Exposure to high noise levels harms the hair cells in the cochlea and damages hearing, with very short exposure to extremely loud sounds and longer exposure to loud sounds being harmful. In the UK regulations require compulsory action at 85 dB (L_{EPd} or daily noise exposure level) and peak SPL of 137 Pa (L_{Cpeak}). Those values are, respectively, roughly equivalent to the noise from a diesel lorry travelling at 40 mph about 15 m away and a jet aircraft 50 m away.

Exposure to potentially harmful levels of sound is usually a result of specific workplace activities, such as working in factories and foundries, or with pneumatic drills or machine tools. Leisure activities, including loud music performances, can also compromise hearing.

Long-term exposure to intrusive noise, particularly in dwellings, can also cause health problems. Initially, noise can make it difficult to get to sleep or produce sleep disturbance, but persistent exposure to noise can result in sleep deprivation, loss of appetite and psychological stress.

Sound and nuisance

Often, noise reaching a space from outside the building or from an adjoining space will not be injurious to health, but will still be a nuisance. Intrusive noise can be distracting, interrupting activities and disturbing concentration. In schools and offices intrusive noise can produce noisy spaces where task concentration or learning is disrupted.

In dwellings, traffic noise, the noise of the neighbour's television through the walls or footsteps on an upper floor, can all diminish quality of life. Problems are more likely in terraces, flats, maisonettes, hotels and halls of residence. Reverberation in circulation spaces, which typically have hard surfaces, increases the disturbance produced by traffic between occupancies.

The impact sound of rain on roofs (particularly those with metal weatherproofing) can also be intrusive.

Sound and privacy

Another aspect of acoustic privacy is the desire not to be heard: this is mainly a consideration in domestic contexts, where the fact that certain human functions are audible in other spaces in a dwelling can cause embarrassment. However, it can also be an issue in open-plan offices.

Sound and communication

Speech is a vital means of communication, so the inability to hear and understand speech is problematic. There are two main considerations:

* Ambient noise – If ambient noise levels are too high then normal conversation and communication will not be possible. The maximum acceptable level of ambient noise for normal speech is 57 dB (L_{Aeq}) where speaker and listener are 1 m apart, but only 39 dB at a distance of 8 m. Offices will typically have ambient noise levels of 45 dB, while a light-engineering workshop will have noise levels of 45–55 dB.
* Reverberation within the space – In some contexts, such as concert halls, long reverberation is desirable, but where speech is important (e.g. in classrooms and lecture theatres) reverberation reduces the intelligibility of speech as the reflected sound muffles the original sound.

Creating spaces which enable clear communication is important for people with a hearing impairment, because overbearing background noise and reverberation will make communication more difficult.

However, in large open-plan offices, intelligibility of speech over a distance is undesirable, as it gives problems with privacy. This can be addressed by siting ventilation outlets so that the noise from the ventilation or conditioning system forms background noise, reducing the audibility of conversations from a distance. Such a set-up will require coordination between the designer and the building services engineer.

Controlling intrusive sound

To reduce intrusive sound, the transmission of sound through the building fabric must be attenuated. That involves applying the basic principles of sound control to the building elements, as well as reducing sound transmission at perimeters and junctions with other elements.

The principles of controlling sound transmission through the fabric

There are four main physical attributes of the building fabric to consider: mass, isolation, rigidity and absorption.

Mass

A sound wave has a finite amount of energy, which is gradually used up as it travels and sets new molecules vibrating. When a sound wave travels through a dense material much of its energy will be used to set the molecules of the material in motion. If there is enough mass all the energy of the sound wave will be absorbed and converted to thermal energy. The amount of mass required to absorb a sound wave depends on the energy of the sound wave: more energy requires more mass. For instance, an impact on a dense floor will produce less sound than the same impact on a lightweight floor.

The use of dense fabric is a common method of reducing airborne and impact sound transmission, and is relatively easy when using masonry or concrete construction. In lightweight framed constructions mass can be increased by adding layers of dense plank plasterboard. As rule of thumb, doubling the mass of a building element will result in a 6 dB reduction in sound transmission. Note that the mass of the fabric is usually expressed as the mass per unit of surface area (kg/m^2), rather than a density value.

Isolation

Building elements usually consist of several different layers. Sound travels less easily through elements whose layers are isolated from each other. This principle is employed in timber-framed partitions and party walls, which have two separate frames, usually lined with acoustic insulation.

In floors, an independent ceiling installed beneath a party floor will reduce transmission by isolating the wearing surface from the ceiling. Where space constraints preclude an independent ceiling, resilient bars can be fitted to the underside of the floor structure to eliminate a direct transmission path between floor and ceiling. At openings, double- or triple-glazed units contain air cavities which reduce sound transmission as well as heat loss.

Any structural connectors required to link parts of the construction (such as wall ties in masonry cavity walls) will increase sound transmission. There is also the possibility of flanking transmission if the cavity is linked to other cavities, so giving a path for airborne sound transmission.

Rigidity

Sound transmission is affected by the rigidity – or **dynamic stiffness** – of materials within building elements: those with low dynamic stiffness transmit less sound and can be used for noise reduction systems. A typical example is the resilient layer used in floating floor constructions, which 'gives' slightly on impact, reducing the transmission of impact sound from the wearing surface. Table 4.3 shows the dynamic thickness for materials commonly used as resilient layers: a dynamic stiffness of 10–50 MN/m³ is usually recommended.

TABLE 4.4 RECOMMENDED REVERBERATION TIMES

MATERIAL	DYNAMIC STIFFNESS, s' (MN/m³)
Mineral fibre boards (10 mm)[1]	20
Mineral fibre boards (20 mm)[1]	10
Granulated cork mats (7 mm)[1]	150
Expanded polystyrene[2]	10-40

SOURCES:
[1] GÖSELE K AND SCHRÖDER E. 2012. *SOUND INSULATION* IN MÜLLER G AND MÖSER M *HANDBOOK OF ENGINEERING ACOUSTICS*, SPRINGER SCIENCE & BUSINESS MEDIA.
[2] PFUNDSTEIN M, GELLERT R, SPITZNER M, RUDOLPHI A. 2008. *INSULATING MATERIALS: PRINCIPLES, MATERIALS, APPLICATIONS.*

Acoustic requirements must be balanced with other performance requirements, such as structural strength, in components such as wall ties, and also long-term stability in resilient layers in floating floor constructions.

Absorption

Porous materials, such as fibrous insulation or open-faced tiles, absorb sound within their air spaces as friction between the vibrating air and the fibrous material dissipates the sound energy and converts it to heat. Some of the sound energy also goes to moving the fibres.

Fibrous insulation is used widely to absorb sound, most commonly in framed constructions which do not have a great deal of mass. Absorptive materials are also used on room surfaces to absorb sound and reduce reverberation (see below: Controlling reverberation p. 114).

Controlling sound transmission in practice

Most constructions derive their acoustic performance from several attributes. This may be illustrated by considering the redevelopment of a large terrace house into multiple flats, in which a floor that was originally an internal floor becomes a separating floor between two flats. As a result, airborne transmission of sound between flats has to be addressed, as well as impact sound transmission from the upper flat to the lower flat. Because this is an existing building, which may be supposed to have a timber joist floor, relying on mass to prevent sound transmission is not an option. A more complex solution is required: one that relies upon isolation, stiffness and absorption.

The typical solution (see Figure 4–07) will involve a floating floor construction, with a resilient layer to reduce impact sound transmission by separating the wearing surface from the timber joists. The resilient layer will be formed of compressible foam, which is either installed across the whole floor, often on the original floor surface, as shown, or along the top of the existing ceiling joists. Fibrous insulation fitted between the joists will improve the acoustic absorption of the floor and dissipate much of the energy in the sound waves.

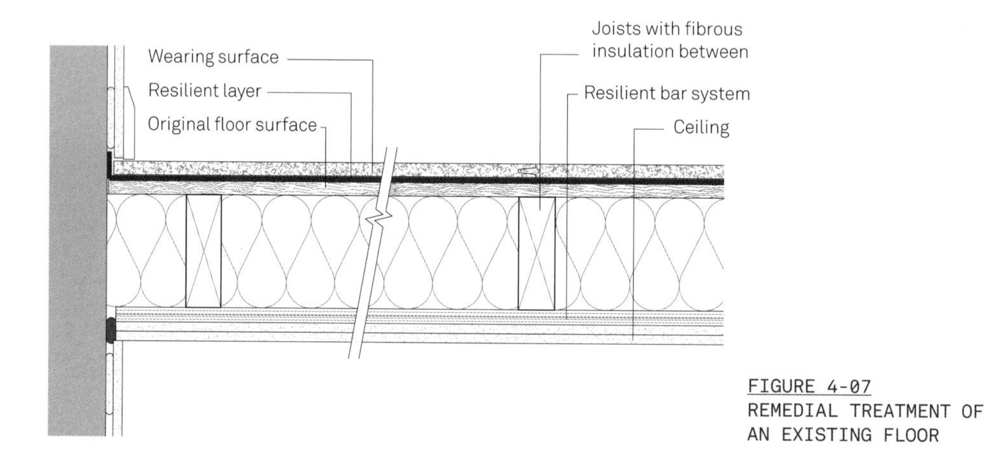

FIGURE 4-07
REMEDIAL TREATMENT OF AN EXISTING FLOOR

The final part of the construction is the ceiling, which needs to be isolated from the joists, The best solution is to construct a separate ceiling which is not directly connected to the joists, but if space is constrained a resilient bar system can be used to minimise contact and sound transmission. To obtain the intended performance the floor must be designed to avoid flanking sound transmission, and services which penetrate the floor must be properly isolated.

Edges, junctions and perimeters

For any sound control solution to be effective the junctions between the treated element and other parts of the building also need to be addressed:

* Where flanking transmission would be likely, for example in a cavity party wall, the flanking elements must be designed to reduce sound transmission. That may require denser construction and the sealing of linked cavities.
* Air gaps at perimeters must be adequately sealed.

* On floating floors, the wearing surface should be separated from the perimeter walls to avoid flanking transmission.
* Service ducts and pipes can be paths for sound transmission, either through gaps where they penetrate walls and floors, or by transmitting sound directly. Services penetrations should be filled with flexible sealants and services themselves protected by acoustically insulated partitions.

Controlling noise from outside

The principles for controlling noise transmission into a building from outside are identical to those for controlling transmission within a building, but with three additional considerations.

Vents

The vents and openings that provide fresh air for the building occupants (see chapter 2: Air for health and comfort p. 53) provide ideal paths for airborne sound transmission. Where external noise is likely to be a problem acoustic vents should be used. These are lined with acoustic foam to absorb sound, or have internal baffles which attenuate sound by setting up interference patterns.

Layout

Where there is a significant noise source outside the building it may be possible to reduce transmission by siting vents and windows away from the noise source. Wherever possible, noise-generating activities on a site should be separated from noise-sensitive ones.

Impact

The impact sound produced by rain on flat roofs can be treated by using thick fibrous tiles in a suspended ceiling to absorb sound. One of the most effective solutions, using a green roof (which provides isolation and absorption), will only be practicable on a limited number of projects.

NOISY NEIGHBOURS

Intrusive noise is hardly a new phenomenon: in the first century the satirist Juvenal complained about the noise of wagons passing along the streets at night (they were banned from Rome during daylight hours).

Measures to control noise have often focused on limiting the amount of noise produced. In early modern London, the 'Lawes of the Market' forbade noisy disturbances at night and limited the hours of noisy trades ('no hammar man, [such] as a Smith, a Pewterer, a Founder, and all Artificers making great sound, shall not work after the hours of nine in the night, nor afore the houre of four in the morning').

By the 18th century the possibilities of limiting noise through building design were increasingly considered: the church of St Mary le Strand in central London was built with false window openings on the ground floor to 'keep out the Noises from the Street'.

That same principle was used more recently for the Byker Wall (a 1970s housing development in the Byker district of Newcastle upon Tyne, England). The towering north face, which was expected to overlook an urban motorway, was designed with only a few, small windows, while the southern, sheltered face has larger windows, walkways and balconies.

Controlling reverberation

The optimum reverberation time required to ensure intelligibility of speech in a room (see above: Sound and communication p. 110) depends on its function and size, with smaller rooms needing lower reverberation times than larger ones. Table 4.4 shows recommended reverberation times for a range of spaces.

TABLE 4.4 RECOMMENDED REVERBERATION TIMES

INTERNAL SPACE	REVERBERATION TIME, IN SECONDS
Private office	0.6–0.8
Open plan office	0.8-1.2
Secondary school classroom	<0.8
Primary school classroom	<0.6
Atrium	1.5–2.0
Restaurant	0.8–1.2

SOURCE: CLARKE SAUNDERS ACOUSTIC CONSULTANT

Where the predicted reverberation time for a room is too high, it can be reduced by selecting different wall and floor coverings and by using acoustic ceilings, wall absorbers and acoustic baffles.

The effect of the acoustic characteristics of surfaces can be seen by examining the reverberation time for an open-plan office which measures 6 m by 10 m, with a height of 3 m (volume, V: 180 m³), and with surface absorption (α) characteristics (for the 1000 Hz frequency band) as follows:

TABLE 4.5

ELEMENT	AREA (m²)	α	Aα
Floor (linoleum)	60	0.05	3.0
Ceiling (plasterboard)	60	0.05	3.0
Walls (plastered masonry)	80	0.02	1.6
Windows	16	0.04	0.64
A_T			8.24

Using the formula for the reverberation time of:

$R_{T60} = 0.16V/A_T$

The reverberation time for the room is:

$R_{T60} = 0.16 \times 180/8.24$

$R_{T60} = 3.5$ s

That reverberation time is excessive for an open-plan office so it must be reduced. One method would be to install absorptive panels (absorption coefficient, 0.95) to the surfaces of the smaller walls, giving revised surface characteristics as follows:

TABLE 4.6

ELEMENT	AREA (m^2)	α	Aα
Floor (linoleum)	60	0.05	3.0
Ceiling (plasterboard)	60	0.05	3.0
Walls (plastered masonry)	44	0.02	0.88
Wall (absorptive panels)	36	0.75	27.0
Windows	16	0.04	0.64
A_T			34.52

The reverberation time is now:

R_{T60} = 0.16 x 180/34.52

R_{T60} = 0.83 s

That figure is within the range of acceptable reverberation times for an open-plan office. If this were a live project the reverberation times would be calculated for a range of frequency bands.

Sound in the bigger picture

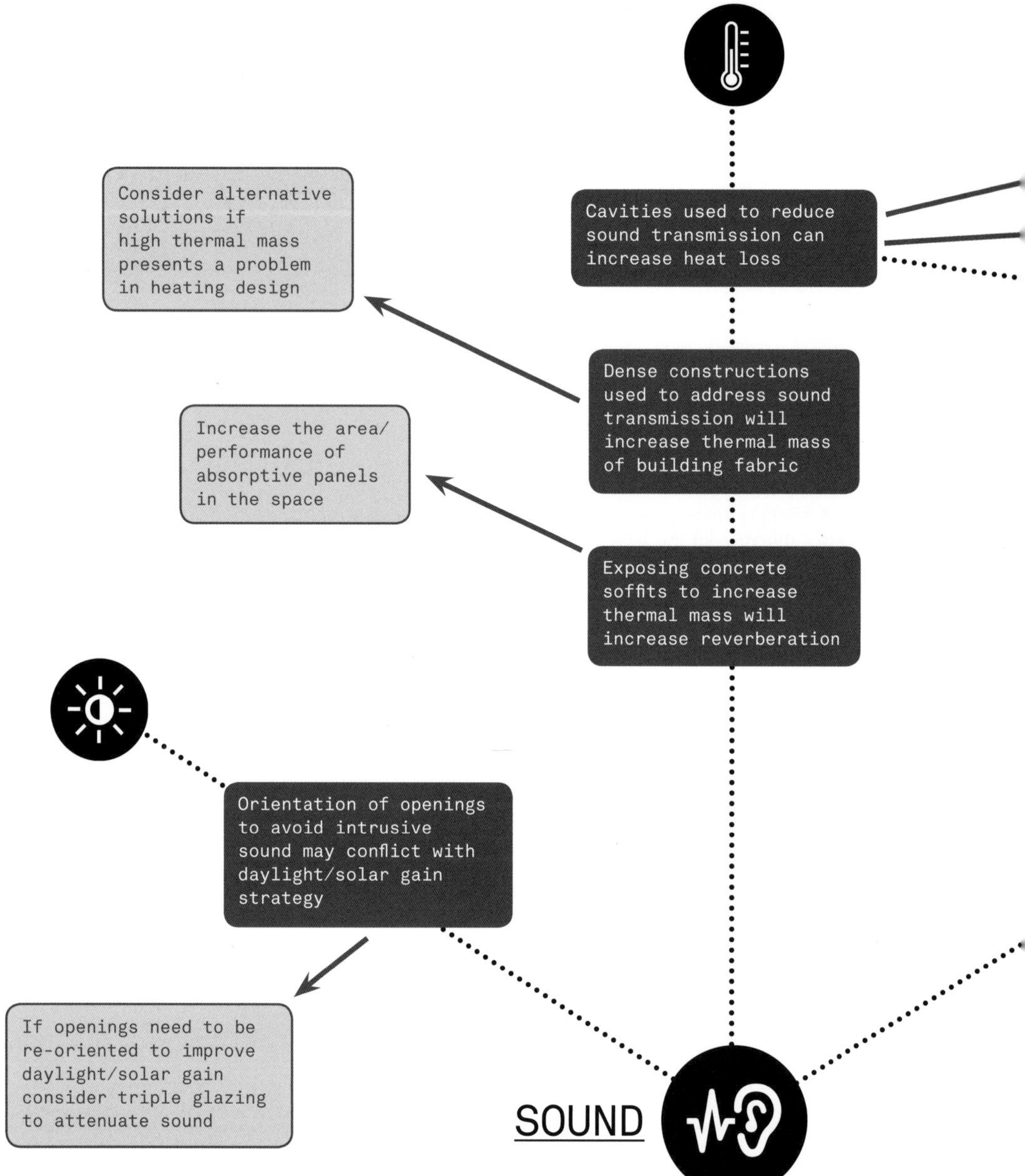

Cavities used to reduce sound transmission can increase heat loss

Dense constructions used to address sound transmission will increase thermal mass of building fabric

Consider alternative solutions if high thermal mass presents a problem in heating design

Exposing concrete soffits to increase thermal mass will increase reverberation

Increase the area/performance of absorptive panels in the space

Orientation of openings to avoid intrusive sound may conflict with daylight/solar gain strategy

If openings need to be re-oriented to improve daylight/solar gain consider triple glazing to attenuate sound

SOUND

INTERACTIONS WITH SOUND

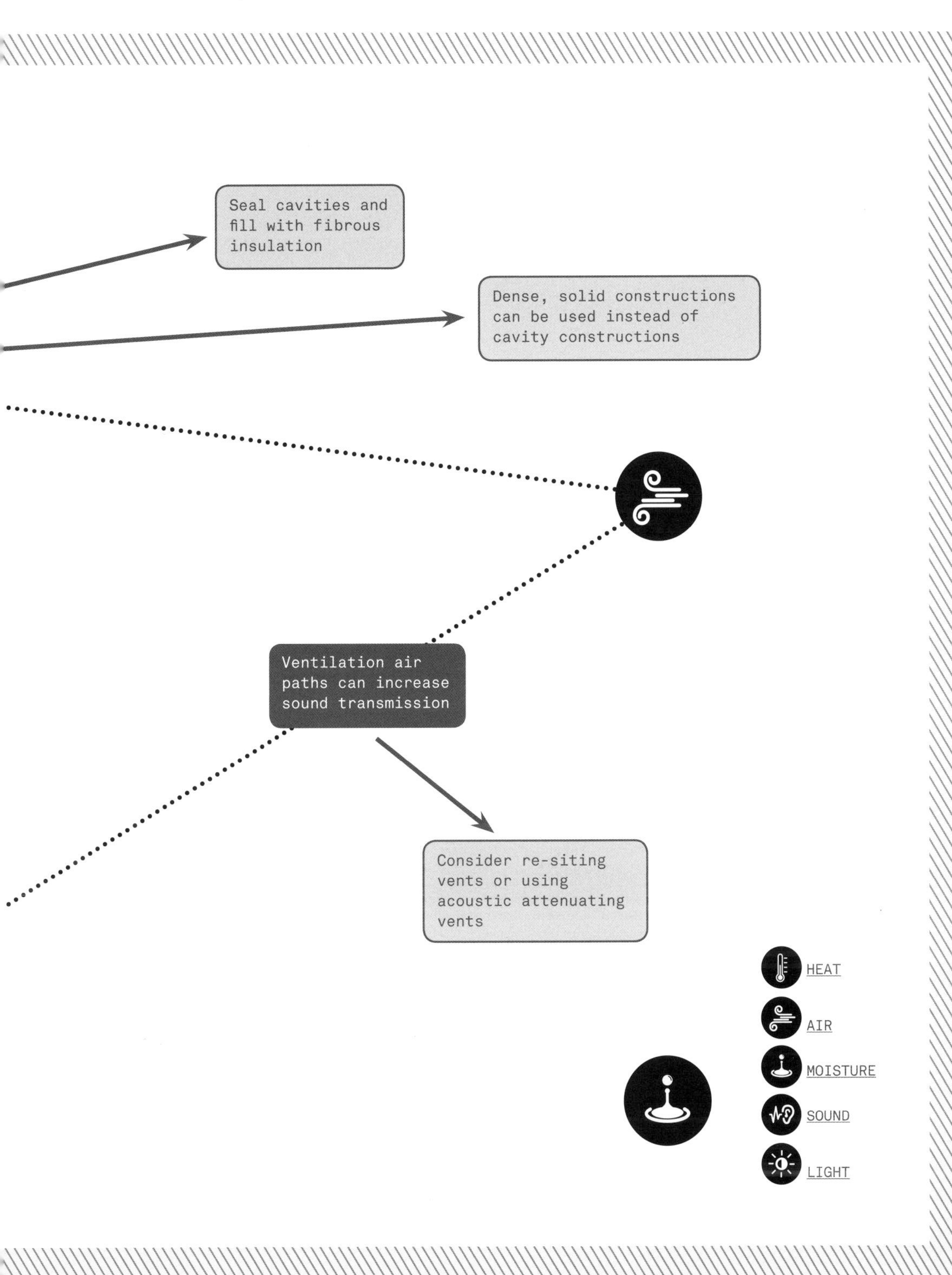
Seal cavities and fill with fibrous insulation
Dense, solid constructions can be used instead of cavity constructions
Ventilation air paths can increase sound transmission
Consider re-siting vents or using acoustic attenuating vents
HEAT
AIR
MOISTURE
SOUND
LIGHT

05

Light

DEFINITION:
LIGHT: ELECTROMAGNETIC RADIATION WITH WAVELENGTHS OF 380–780 nm, TO WHICH THE HUMAN EYE IS SENSITIVE.

The earth's atmosphere is thick with electromagnetic radiation, emitted by sources such as stars, mobile phones, people and buildings. Human beings can only directly detect one narrow band of radiation – the band we recognise as running through the colours of the rainbow from red to violet: visible light. On either side of the band of visible light are the infrared and ultraviolet wavelengths: although we cannot see either of those, we can feel the warming effect of infrared and experience the harmful effects of ultraviolet (e.g. the 'healthy tan' which is skin damage).

The human response to, and interaction with, light is complex: at its simplest, we need light to see, but the colour and intensity of the light we see affects our perception and even our mood (the light on a summer's morning and a winter afternoon feel different). Although we have an extensive range of artificial light sources available there is a growing understanding that daylight provides much more than illumination (see below: Physiological effects of light p. 125). Moreover, the need to minimise energy consumption in buildings makes it all the more important to make good use of daylight.

In order to understand how to design buildings which are suitably illuminated and energy efficient, we must first understand the fundamental nature of light, then consider how human beings perceive light and also the wider effects of light on humans. We can then examine the principles of governing the provision of light in buildings and how those principles may be applied to daylight and to artificial light. The effect of solar gain on the thermal performance of buildings is considered in chapter 1 Solar gain p. 31.

Fundamentals of light

Infrared, visible and ultraviolet light are all forms of electromagnetic radiation (EMR), generated at an atomic level by the interaction of electrical and magnetic forces. EMR consists of particles (photons) which behave like waves and, unlike sound, can travel through a vacuum. In a vacuum, EMR moves at a speed of 3×10^8 m/s (i.e. the speed of light), but more slowly through media such as gases, glass (2×10^8 m/s) or water (2.3×10^8 m/s).

The wavelength of EMR ranges from around 2000 m for long wave radio to 1.0×10^{-12} m for gamma rays. Figure 5–01 shows the range of the EM spectrum, together with the common uses of the various parts of it (the scale is logarithmic: each gradation being ten times smaller than the one below). Visible light has wavelengths between 780 nm (red) and 380 nm (violet), while the thermal radiation discussed in chapter 1 (see: Radiation) occurs at around 100 μm.

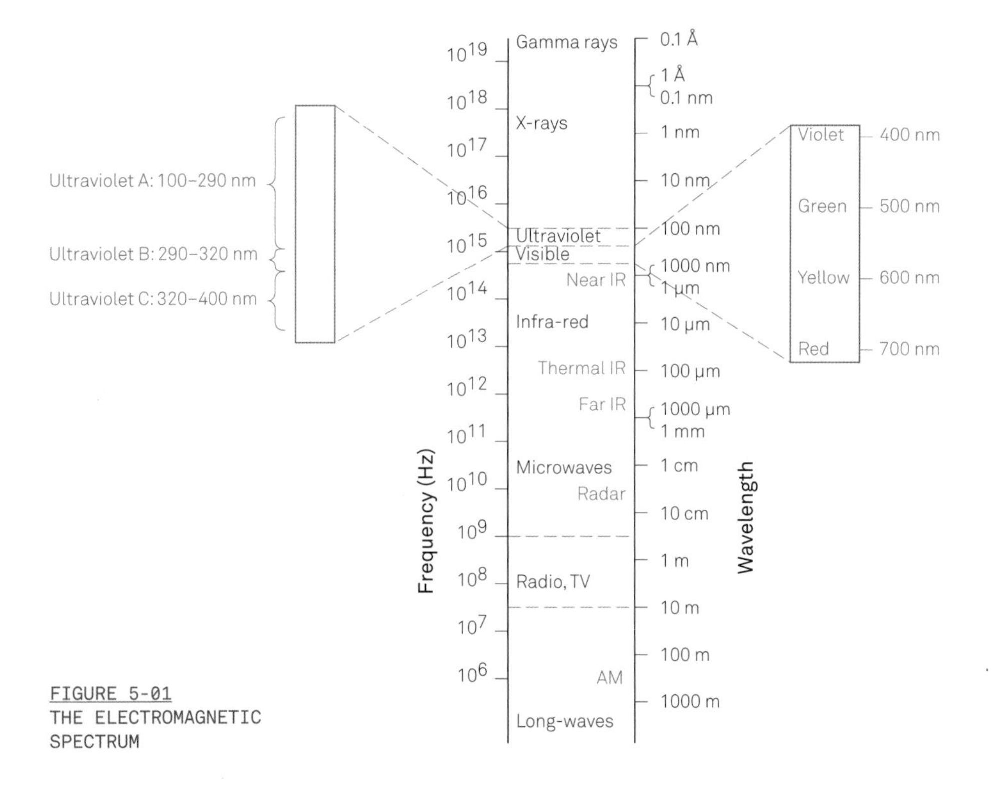

FIGURE 5-01
THE ELECTROMAGNETIC SPECTRUM

Light and materials

The interaction of light with materials depends on the wavelength of the radiation and the characteristics of the material. Gases and liquids will allow some light to pass unchanged (although at a slower speed), but will absorb or scatter other wavelengths. For example, the blue of the sky is a result of shorter wavelength blue light from the sun being scattered in the atmosphere and so reaching us from a wider angle. Solid materials will absorb, reflect or transmit light:

* Absorption – occurs where the frequency of the light photons is similar to the natural frequency of the electrons in the material. Absorption raises the temperature of the material (see chapter 1: Solar gain p. 31).
* Reflection – occurs where the frequencies of the light photons and electrons are dissimilar. Reflectance is the proportion of incident light which a surface returns: 1 represents complete reflection, 0 complete absorption.
* Transmission – occurs when the atoms that make up the material contain electrons that do not interact with the photons. Photons from the visible part of the spectrum (wavelengths of 380–780 nm) do not interact with electrons of glass molecules, and therefore are transmitted through the material. However, photons from the ultraviolet part of the spectrum (wavelengths of 10–380 nm) can interact, and are absorbed by standard glass.

Light from a source commonly consists of a range of wavelengths; when it strikes a surface some wavelengths will be absorbed and some reflected (or transmitted). It is only the reflected wavelengths which reach the human eye and are recognised as colour. For example, an object which reflects green light but absorbs all other light will appear green.

The sun

The sun is the main natural source of light on earth: it emits a wide spectrum of EMR, some of which is absorbed by the atmosphere, so that at the earth's surface the radiation is composed of 53% infrared, 44% visible light, and 3% ultraviolet. Because the earth rotates around the sun and rotates on its own axis (which is 23.4° from vertical) the angle of the earth to the sun changes constantly, resulting in variations in the length of day, the angle of the sun in the sky and the intensity of solar radiation.

The change in day length depends on latitude: at the equator there is no significant variation, but the nearer a site is to the poles, the greater the variation in day length between winter and summer, until at the Arctic and Antarctic circles there is perpetual daylight during the summer and no daylight during the winter. Figure 5–02 shows the annual variation in daylight hours for three locations.

The change in the length of day is linked to the position of the sun in the sky. Figure 5–03 shows the sun position through the year for latitude 55° north and midsummer and midwinter. In the summer, the sun rises and sets over a wider range, and is higher in the sky at noon, so solar radiation is more intense.

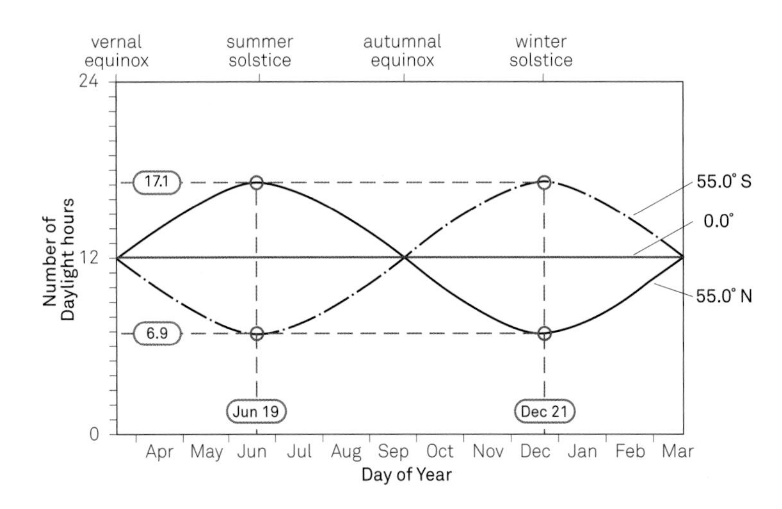

FIGURE 5-02
VARIATION IN DAYLIGHT HOURS THROUGH THE YEAR FOR THREE LOCATIONS

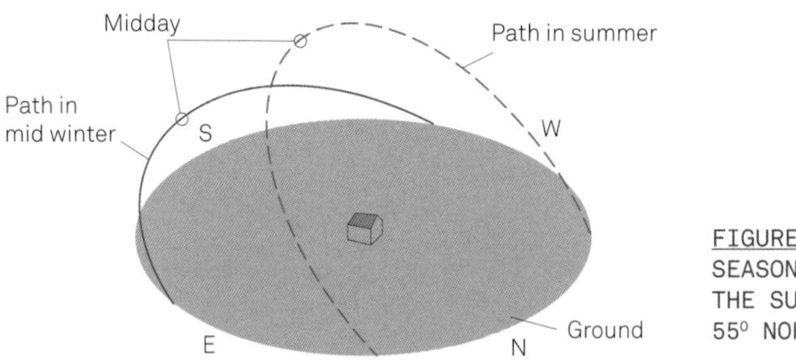

FIGURE 5-03
SEASONAL CHANGES IN THE SUN'S PATH FOR 55° NORTH

Measuring light

In defining light there are two main characteristics to consider: the intensity (or brightness) and colour.

Intensity

The strength of a light source, its **luminous intensity**, is measured in **candela** (cd) (roughly equivalent to the light given out by a candle flame). Luminous intensity is the power of visible light emitted for a solid angle of one **steradian** (an angle which encompasses roughly 1/12.6 of the surface of a sphere). One candela is 1/683 watt per steradian. In practice the contribution of each wavelength is weighted according to the sensitivity of the eye to different wavelengths.

The total amount of visible light emitted by a source, the **luminous flux**, is measured in **lumens** (lm). A light source radiating 1 cd in all directions has a luminous flux of 12.6 lm, while a source radiating with the same intensity but in a hemisphere would have a luminous flux of 6.3 lm. The light received by a surface is expressed as the **illuminance**, measured in **lux** (lx), which has the units lumens/m^2.

The amount of light a surface receives from a source decreases as the distance between them increases (it is inversely proportional to the square of the distance between source and surface). As Figure 5–04 shows, a source producing an illuminance of 160 lux on a surface 1 m away will produce an illuminance of 40 lux on a surface 2 m away, and 10 lux on one 4 m away. That decrease affects the performance of all artificial light sources. (The same principle applies to light from the sun, but the earth's distance from the sun is so great that any changes produced by variations in the earth's orbit are negligible for our purposes.)

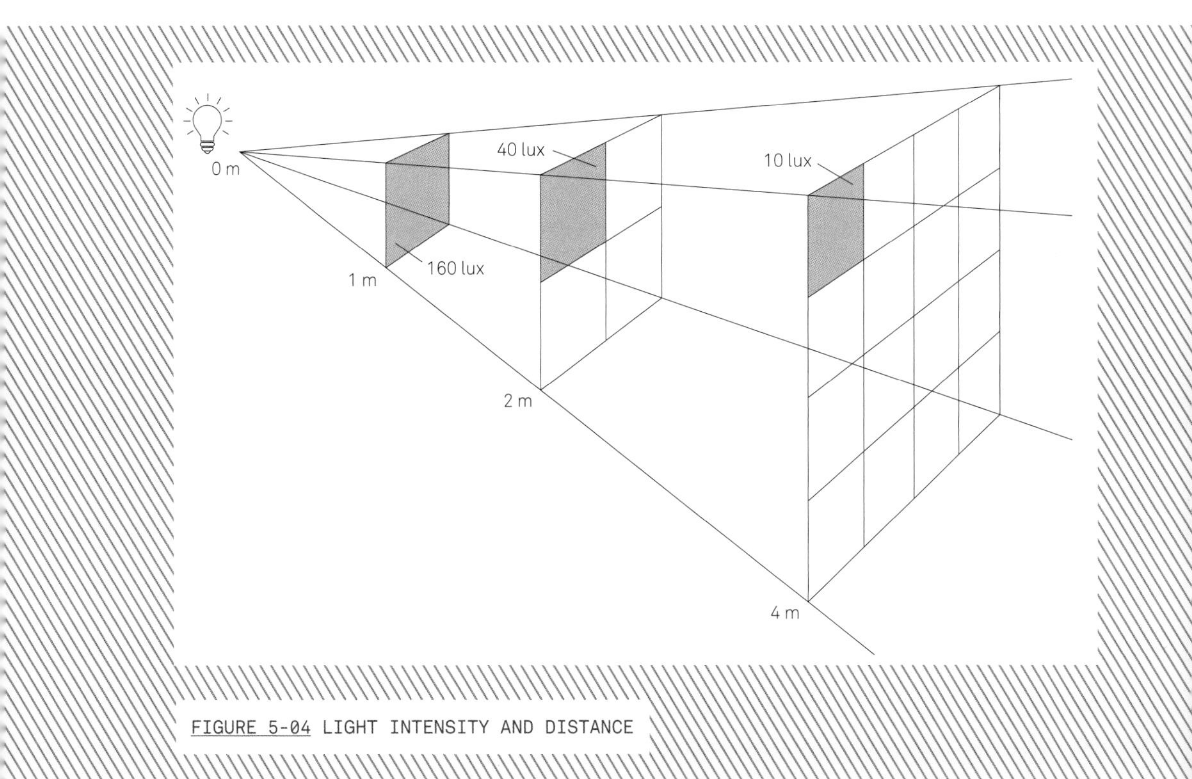

FIGURE 5-04 LIGHT INTENSITY AND DISTANCE

Colour

An accurate description of colour has three components: the colour of the light emitted by sources; the colour of the incident surface; and the 'rendering' of colour (i.e. how accurately does the colour appear under a specific light source).

The majority of light sources emit light in a band of wavelengths, which is determined by the temperature of the source. Colder sources, such as candles and fires, produce predominantly red and orange light, while hotter source, such as the sun, produce bluer light.

The wavelengths of light produced by a particular source can be matched with those from an ideal light source (known as a black body radiator) at a given temperature. That gives the colour temperature of the light source. Figure 5–05 shows the colour temperatures for common light sources. The colour temperature can also be used to describe light which is not emitted by hot sources, such as light-emitting diodes (LEDs).

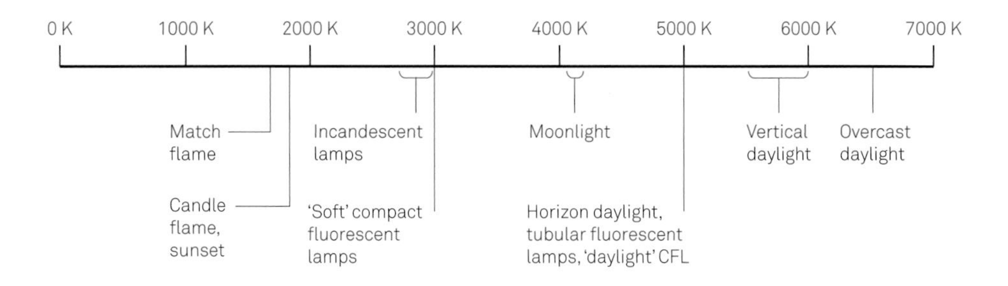

FIGURE 5-05 TYPICAL COLOUR TEMPERATURES OF LIGHT SOURCES

There have been numerous attempts to create systems consistently defining colours (something which even a quick glance at a paint chart shows can be difficult). Most systems define a 'colour space' (a mathematical attempt to describe the relationships between colours), then add a measure of the intensity of the colour (saturation) and sometimes its lightness. The main systems in use today are:

* Munsell – considers hue, based on the dominant part of the spectrum (red, yellow, green, blue, purple), chroma (the strength of the colour from neutral grey (0) to saturation), and value, between perfect black (0) and perfect white (10).
* NCS (Natural Colour System) – uses a colour space with white, black, red, yellow, green and blue, together with an expression of colour intensity (chromaticness) and a black–white axis.
* DIN – uses a colour space defined by hue, saturation and darkness.
* RAL – a system of 1688 colours defined on hue, lightness and chroma.
* BS 5252 – a system originally designed for coordinating colours within buildings. It comprises 237 surface colours, defined with a hue (numbered 00 to 24), greyness (letters A to E) and weight (an additional number). So 00 A 01 is a colour described as ash grey/oyster. BS 4800 contains a table which aligns BS, NCS and Munsell.
* CIE L*a*b* – uses three parameters: L* refers to lightness, while a* and b* denote colour. The system is designed to enable differences between colours to be expressed, as well as describing colours.

The final aspect of colour is colour rendering: how a colour appears under a light source and whether colours 'look right'. The commonest measure is the colour rendering index (CRI), which compares the appearance of standard colour samples illuminated by the light source with their appearance illuminated by a standard light source. The maximum CRI is 100, which indicates a light source identical to standardised daylight (see: Colour rendering p. 127).

Light, people and buildings

Human interaction with light is complex, encompassing physical health, vision and physiology, as well as less measurable effects on mood and emotions.

Physical health

Ultraviolet (UV) light has higher energy levels than visible light (see above: The sun and Figure 5–01 p. 120) and affects the skin and eyes. Light in the UV part of the spectrum encompasses a range of wavelengths that have different physical effects, and is therefore categorised by type; the best known being UVA (315–380 nm) and UVB (280–315 nm). In animals, moderate exposure to UVA increases the production of melanin skin pigment, which protects against UVA and UVB, but high exposure to UVA can result in melanoma (skin cancer).

UVB exposure is vital for humans to maintain healthy levels of vitamin D; however, overexposure to UVB causes sunburn and direct damage to DNA, leading to skin cancer, and can also damage the eyes.

Less well known is UVC, which is also harmful, but the UVC wavelengths of sunlight are filtered out by the earth's atmosphere, so there is no natural exposure.

Vision

The primary human perception of light is through the eye, with stimuli relayed to the brain for processing. The retina at the back of the eye is covered with photoreceptors – light sensitive cells known as rods and cones. Rods are sensitive to the presence of light and primarily detect motion, while cones are sensitive to colour. When light levels are low, vision is provided mainly by rods, but at the higher levels required in buildings the cones predominate. Roughly 5% of people are colour blind in the red/green region, which should be taken into account when designing colour schemes.

Physiological effects of light

A third type of photoreceptor, ganglion cells, send signals of light and dark directly to the hypothalamus, the part of the brain that regulates many physiological functions. Those signals help to regulate the circadian systems which control sleep patterns, changes in body core temperature and some hormone secretion. Prolonged absence of daylight, particularly in the morning, disrupts the circadian systems, leading to poor sleep quality and depression. Low light levels in winter can result in seasonal affective disorder (SAD), which is characterised by depression, lack of energy and increased appetite.

Those physiological effects must be taken into consideration – especially for buildings which people are likely to occupy for long periods, such as hospitals and carehomes.

Psychological effects of light

Light can subtly affect mood and perception. We think of reds and oranges as warm and blue as cold: which is at odds with the physical reality (see above: Colour p. 123). We feel the difference in quality of light between a summer's morning and a winter afternoon. This is probably the subtlest interface between physics and the human experience of buildings, and as such, the most difficult to codify.

Some of these psychological effects can be addressed by effective use of daylighting. The design of artificial lighting should also address these.

Lighting buildings

This section sets out the principal requirements for establishing good lighting conditions and considers how they may be provided using daylight and artificial light.[1]

Principles of good lighting

Lighting conditions should enable the occupants of a building to carry out their activities safely and comfortably, taking into account their physiological and psychological needs. The lighting requirement will vary according to the function of a space and the tasks carried out within it.[2] Four key criteria for achieving good lighting conditions are: **illuminance**, **modelling** (the balance of direct and diffuse light), **colour rendering** and **visual contrast** of surfaces.

Illuminance

There must be sufficient light for the activities that are to be undertaken in the space. Table 5.1 gives broad guidelines on illuminance for common activities. Illuminance should be measured at the work plane which, for some applications, may well be vertical (e.g. warehouses or libraries). Areas where specific tasks are carried out may require more illumination than is needed for the whole space (e.g. a reception desk in a foyer), in which case task-specific lighting will be needed. The area immediately surrounding a task area (0.5 m beyond) should be illuminated as the final column of table 5.1. The background area (up to a further 3 m from the task area) should be illuminated to at least 1/3 of the immediate surrounding area.

TABLE 5.1 RECOMMENDED ILLUMINANCE LEVELS

INTERIOR TASKS	LOCATION/FUNCTION	TASK AREA ILLUMINANCE (lux) (lumens/m²)	ILLUMINANCE (lux) OF IMMEDIATE SURROUNDINGS (lumens/m²)
Rarely used interiors, visual tasks confined to movement and casual seeing	Corridors and stores	100	100
Visual tasks that do not require perception of detail	Foyers, entrances	200	150
Moderately easy visual tasks	Libraries, sports halls, lecture theatres, background office lighting	300	200
Moderately difficult visual tasks where colour judgement may be required	Office desks, kitchens, laboratories, retail	500	300
Very difficult visual tasks, where small details have to be perceived	General inspection, electronic assembly, retouching paintwork, cabinet making	1000	500

SOURCE: LG10 DAYLIGHTING - A GUIDE FOR DESIGNERS (2014); THE SLL CODE FOR LIGHTING (2012) (BOTH: SOCIETY OF LIGHT AND LIGHTING).

Light should also be evenly distributed within a space. The distribution of light is affected by the amount of light reflected back from surfaces: the following values of reflectance are recommended:[3]

* Ceilings: 0.7–0.9
* Walls: 0.5–0.8
* Floors: 0.2–0.4.

Differences in illuminance can be used to draw attention to features: for example, in display lighting a ratio of 5:1 between background and feature illuminance provides a definite distinction, while 30:1 provides a dramatic effect.

Too high a level of illuminance within a space can result in glare on surfaces, which is visually disturbing. Poorly positioned window and inadequately shaded luminaires may produce distracting reflections on computer.

Modelling

Objects can be illuminated either by light coming directly from a source (**directional lighting**) or by **diffuse lighting** reflected from multiple surfaces. To help people perceive clearly the shape and form of objects, and to be able to read human faces, there should be directional and diffuse lighting. Overly directional lighting (such as strong sunlight or bright spotlights) produces deep shadows, while diffuse lighting on its own (e.g. translucent rooflights in a high-ceilinged warehouse or retail unit) produces dull visual conditions.

The amount of light on three-dimensional objects is assessed using **cylindrical illuminance**, which is the average illumination on the curved surface of a cylinder. A cylindrical illuminance of 150 lx will be sufficient for the clear perception of human faces.

Colour rendering

Lighting should enable building users to discern accurately the colours of surfaces and objects. Poor colour rendering makes it is more difficult for people with visual impairment to distinguish between surfaces. Accurate colour rendering is more important for some activities (e.g. design workshops) than others (e.g. night clubs).

Visual contrast of surfaces

For people with normal vision, contrasts in colour (hue) and intensity (chroma) of surfaces enable clear perception. However, for partially sighted people, it is the difference in light reflectance values (LRV) between surfaces which is most significant.

LRV, which is not related to colour, ranges from 0 (perfectly absorbing) to 100 (perfectly reflecting). A difference of 30 between the LRVs of adjoining surfaces gives good visual contrast, but a different of 20 is sufficient where surface illumination exceeds 200 lux; a difference of 15 suffices for three-dimensional features such as door furniture.[4] Differences in LRV are most important for distinguishing small objects on a larger background (e.g. a switch plate on a wall). Poor colour rendering will hamper the perception of LVR differentials.

Daylight

Daylight has a value beyond the illumination of spaces and of tasks. In a day-lit room the brightness varies with time, the colours are rendered well and the direction of lighting gives good three-dimensional modelling.[5]

Daylight has two components:

* Direct sunlight, which falls on a surface
* Skylight, the diffuse light produced by scattering of sunlight in the atmosphere

Direct sunlight is beneficial, but to avoid visual or thermal discomfort, it should not fall on people working, or on visual tasks. The orientation of the building and its openings should take account of the direction of sunlight at all seasons (see above: The sun p.121). Where occupants have a reasonable expectation of sunlight a room should receive sunlight for at least a quarter of all daylight hours (5% in winter). In north-facing rooms and dense urban settings the absence of sunlight is acceptable.

Skylight can provide good general illumination and the design of the building should ensure that there is not too great a contrast between illumination inside and outside. As a rule of thumb, daylight will penetrate a room for a distance equal to 2.5 times the window head height, which typically gives a penetration of about 6 m. A dual-aspect building can therefore have a 12 m deep plan and still rely on daylighting.

The extent of daylight penetration can be measured by the 'no-sky line', which is the line at which the sky cannot be seen from the height of the working plane (taken as 0.70 m in offices, 0.85 m in houses); this is illustrated in Figure 5–06.

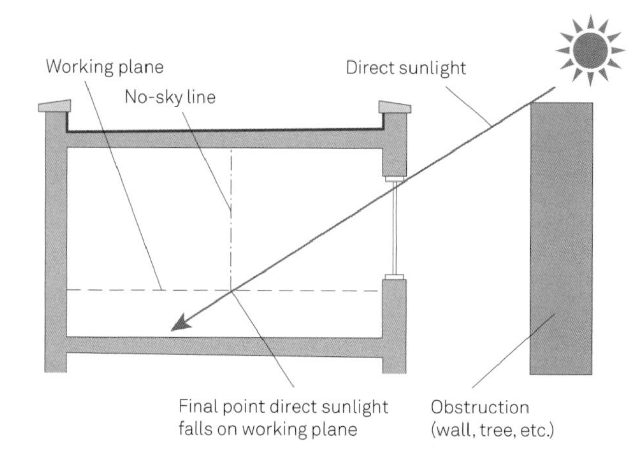

FIGURE 5-06
DAYLIGHT MEASUREMENTS

The amount of daylight a space will receive can be assessed using the average daylight factor (ADF), which expresses the illuminance on an internal surface as a percentage of a standard outside illuminance (see box: Calculating the average daylight factor p. 129). The ADF can be used as a rough guide to the need for artificial light:

* ADF 5% or more – the room will be well lit; some artificial lighting may be required at the start and end of the day
* ADF 2–5% – the room will have a predominantly day lit appearance, but is likely to require some artificial light during the day
* ADF less than 2% – the room will be gloomy and will require artificial light for most of the day

In single-storey buildings, and on the top floors of multistorey buildings, rooflights can be used to provide daylight beyond the range possible for windows. It may also be possible to transmit daylight into the back of a room using a sun pipe – a reflective tube which has a rooflight (usually on a pitched roof) at the upper end and a ceiling-mounted diffuser at the lower end.

The combination of directional sunlight and skylight means that daylight provides good modelling of faces and objects. Daylight gives good colour rendering, although the balance of light in the spectrum will change on both a daily and seasonal basis.

CALCULATING THE AVERAGE DAYLIGHT FACTOR

The ADF on the working plane of a room can be calculated using the equation:

$$ADF = \frac{TA_W\theta}{A(1-R^2)}$$

Where:

T is the diffuse light transmittance of the glazing, including the effects of dirt
Aw is the glazed area of the window in m^2, net of the frame and glazing bars
θ is the angle of the visible sky (degrees), measured from the centre of the window, as shown in Figure 5–07
A is the total area of the ceiling, floor and walls, including windows, in m^2
R is the area-weighted average reflectance of the interior surfaces. A figure of 0.5 can be used as a rough value for a room with a white ceiling and walls of average reflectance

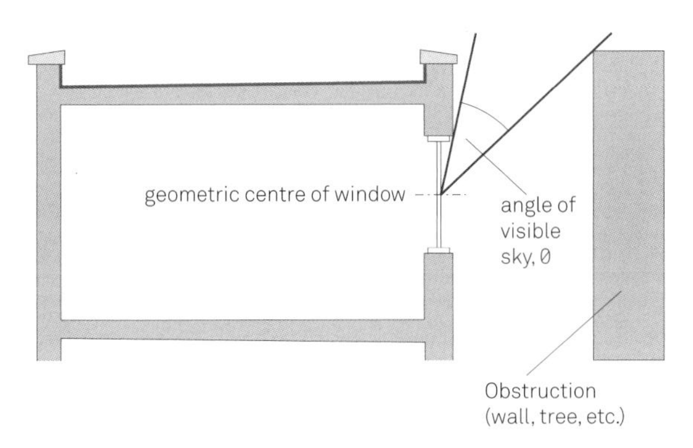

Artificial light

Artificial lighting should provide a visual environment which is:

* Safe – people can move around the space easily
* Efficient – people can carry out their tasks properly and unhindered
* Comfortable – lighting does not strain the eyes through glare, and does not form veiling reflections or flicker

Good lighting design can provide an inspiring and creative visual environment. The three principles outlined above (see: Principles of good lighting p.126) form a good starting point.

A lamp's efficiency is expressed by its luminous efficacy, which is the amount of light emitted for each watt of electricity, measured in lumens per watt (lm/W) (see below: Light sources, p. 131 for typical values). The efficiency may also be quoted as a luminaire luminous efficacy, which takes account of the power absorbed by the luminaire's circuits and the physical effect of the luminaire on light output from the lamp.

Illuminance

The amount of light provided by an individual luminaire will depend on its type and configuration, and the design of shades and reflectors. **Photometric data sheets** for luminaires contain intensity tables which give the intensity of light produced at a range of angles. The intensity values are used to calculate surface illuminance. While lighting calculations are now usually carried out using specialist software, the number of luminaires required for large installations can be estimated using the lumen method.

THE LUMEN METHOD

A rough calculation of the number of lamps required for a space, N, can be carried out using the following equation:

$$N = \frac{E \times A}{F \times MF \times UF}$$

Where:

E is the illuminance level required
A is the area at the working plane, in m^2
F is the average luminous flux from each lamp, in lumens
MF is the maintenance factor, allowing for the reduction in output caused by deterioration and dirt
UF is the utilisation factor, which is the proportion of light from the lamp which will illuminate the working plane. (The utilisation factor depends on the room configuration and the luminaire design.)

For example, an office measuring 20 m x 10 m is to be illuminated to a level of 500 lux, using luminaires containing lamps with an output of 15,000 lumens, a maintenance factor of 0.75 and utilisation factor of 0.5. The number of lamps required is:

$$N = \frac{200 \times 500}{15{,}000 \times 0.75 \times 0.5}$$

$$N = 17.8$$

or, 18 lamps.

Visibility under artificial light

Luminaires should not give overly directional light, which produces hard shadows, nor overly diffuse light, which masks detail. The extent of modelling can be evaluated from the ratio of the **horizontal illuminance** (i.e. the illuminance on a horizontal surface) and the cylindrical illuminance at a point. For environments with a regular array of luminaires a ratio of 0.3–0.6 will give good modelling. Additional directional lighting may be required for visually demanding tasks.

Colour rendering

The colour rendering of light sources in a space must be matched to the activities taking place: as a rule, the more important vision is for an activity the better colour rendering is required. For example, corridors require a CRI of at least 40; plant rooms, 60; hairdressing, 90.

Light sources

There are four main types of artificial light sources used in buildings:

* Incandescent lamps – light is produced by passing an electrical current through a tungsten filament, which makes it glow. The majority of the radiation emitted is infrared. These lamps have a colour temperature of about 2700–3000 K, giving a warm, yellow light. Colour rendering is good. Luminous efficacy: 14–16 lm/W.
* Halogen lamps – these are incandescent lamps with halogen gas (iodine or bromine) in the enclosure to extend the filament life. They are hotter than incandescent lamps, so produce light at a high colour temperature. Luminous efficacy: 20–35 lm/W.
* Fluorescent lamps (including compact fluorescent lamps, CFLs) – an electric current is passed through mercury vapour to ionise the gas, producing photons which make the fluorescent (phosphor) coating on the glass glow. The colour temperature is in the range 2700–6500 K depending on the composition of the phosphor coating. Colour rendering also varies with coating: the CRI can be as low as 50, but daylight fluorescents can approach 100. Luminous efficacy: 50–100 lm/W.
* LEDs – these are semi-conductor chips which emit light when a current is passed through them. They do not have a conventional colour temperature. Colour rendering is improving as the technology develops. They have a long life span (approximately 25,000 hours compared with 1000 for incandescent). Luminous efficacy: 50–90 lm/W.

Light in the bigger picture

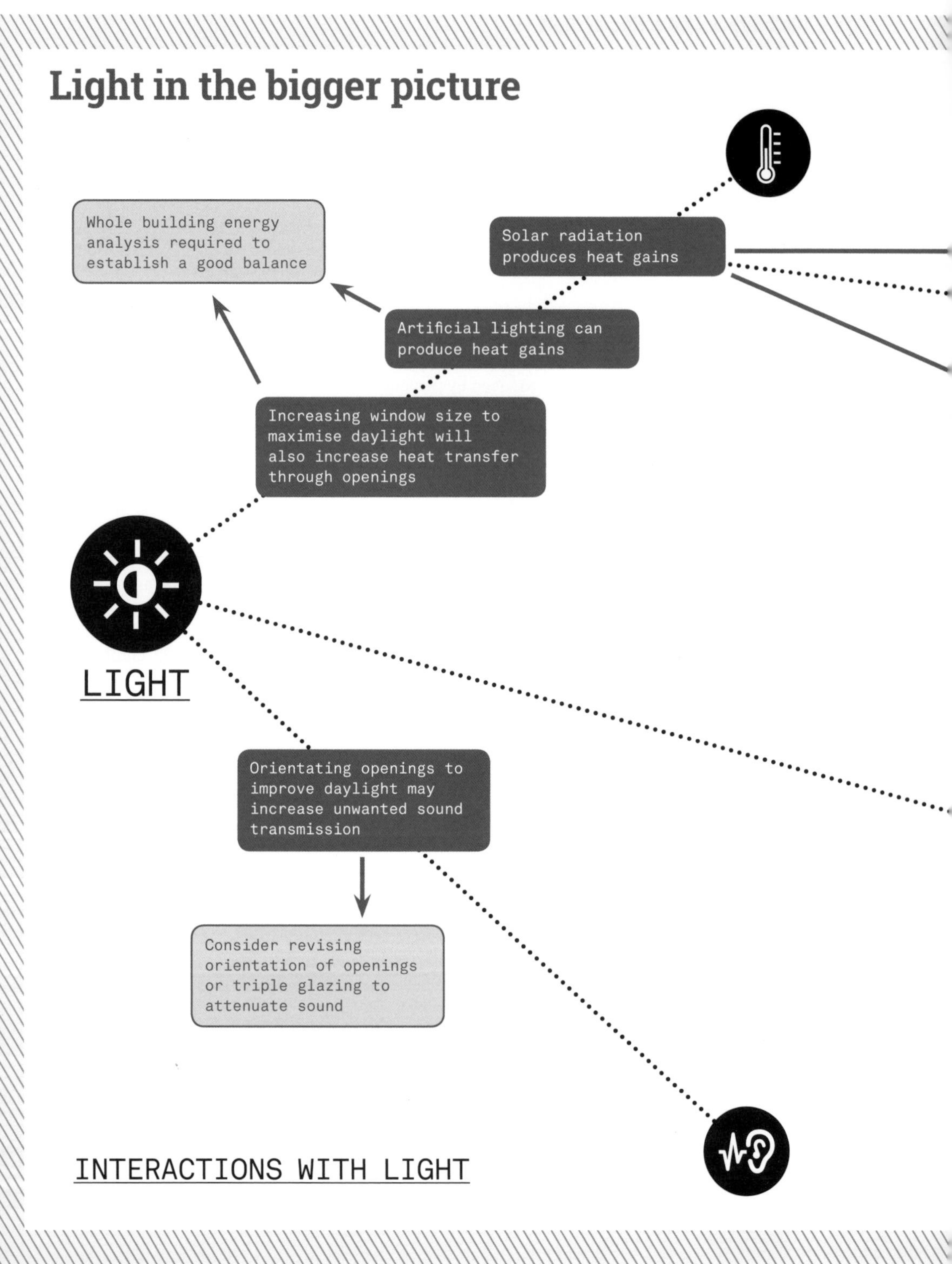

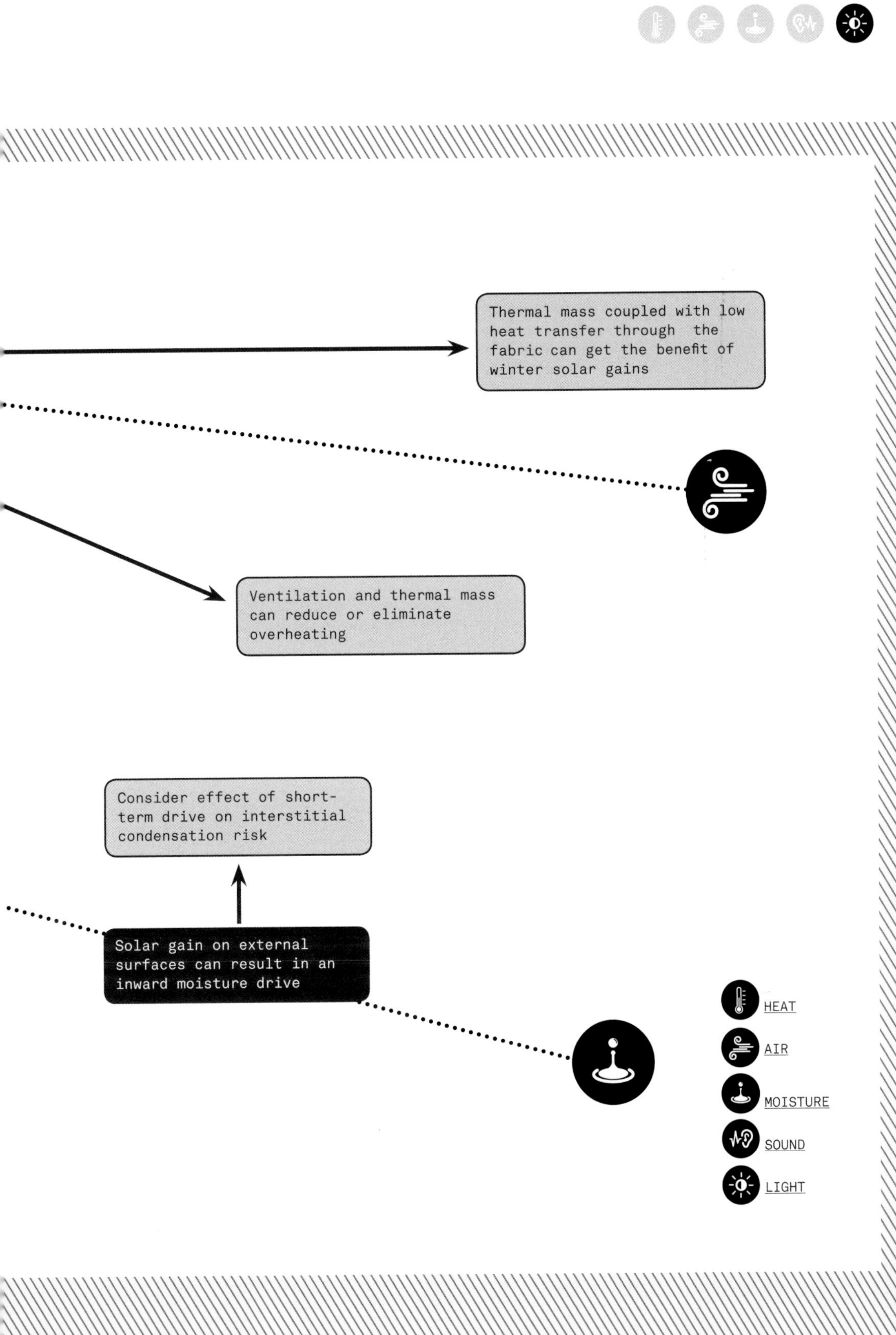
Thermal mass coupled with low heat transfer through the fabric can get the benefit of winter solar gains
Ventilation and thermal mass can reduce or eliminate overheating
Consider effect of short-term drive on interstitial condensation risk
Solar gain on external surfaces can result in an inward moisture drive
HEAT
AIR
MOISTURE
SOUND
LIGHT

Glossary

Absorptivity	a measure of the amount of radiation a surface absorbs, relative to a notional black body. Values range from 0–1.
Admittance	a measure of how rapidly heat will pass between the surface of an element and the interior of the building.
Advection	the transfer of heat by means of movement of a material, for example, air movement cause by differential wind pressure.
Air permeability	a measure of the rate of air infiltration though the external envelope of a building. Expressed as the volume of air passing through one square metre of the building surface per hour, at a pressure difference of 50 Pa ($m^3/m^2.h$ at 50 Pa).
Airborne transmission	the transmission of sound around a building, either directly through gaps and openings, or indirectly as sound waves generate vibrations on one side of building element which then generate sound waves on the other side.
Becquerel (Bq)	the unit of radioactivity, defined as one transformation or decay per second. Named after Henri Becquerel, a pioneer researcher of radioactivity.
Candela (cd)	the measure of luminous intensity. One candela is 1/683 watt per steradian.
Capillary action	the movement of water through narrow spaces in materials, often upward, produced by surface tension in the liquid and the adhesive forces between the liquid and the materials.
Centre pane U-value	the U-value of an insulated glazing unit, measured perpendicularly through the middle of the unit.
Combined method	a simplified method of calculating the U-value of a building element.
Compressive strength	a measure of how a material performs under load. Often quoted as a pressure which will produce a 10% deformation, or a 2% long-term deformation.
Conduction	heat transfer within materials resulting from energy being passed from atoms with high internal energy to those with lower internal energy.
Convection	the movement of atoms in a fluid as a result of variations in density, often produced by temperature differences.
Cylindrical illuminance	a measure of light distribution on cylindrical objects. A value of 150 lx or greater will give lighting which provides good illumination of faces.
Decrement delay	the time taken for heat to pass from one side of the structure to the other.
Decrement factor (f)	the ratio of the internal temperature range (minimum to maximum) to the external temperature range.
Dew point temperature	the temperature at which air at a given vapour pressure would be saturated.
Diffuse lighting	lighting conditions where light reaches surfaces after reflection from several surfaces.
Directional lighting	light reaching a surface directly from a light source. Sunlight provides directional lighting.
Dry bulb temperature	the ambient temperature of air, measured out of direct solar radiation.
Dynamic simulation modelling	a method of analysing the energy performance of buildings which simulates the performance on an hourly basis.
Dynamic stiffness	a measure of the rigidity of materials.
Electromagnetic radiation	energy waves of photons generated by sub-atomic forces.

Emissivity	a measure of the effectiveness of a body as a source of radiation. Expressed as a fraction of the performance of a perfect source of radiation (a 'black body'). Values range from 0–1.
Enthalpy	the total of the sensible and latent heat in the air, which will vary as the air is heated or cooled, or the moisture content changes. Used when calculating the energy involved in humidification and dehumidification.
Equivalent area (vents)	the area of a single opening which would provide the same air flow as the vent, taking account of air turbulence at the vent openings.
Flanking transmission	the transmission of sound through a building element which is adjacent to one being treated.
Free area of vent	the total unobstructed area of a vent.
Gas laws	the laws of physics which describe the behaviour of ideal gases.
Heat	the internal energy possessed by matter.
Heat capacity	the amount of energy required to raise the temperature of a layer or construction by 1 K.
Horizontal illuminance	the illuminance on a horizontal surface.
Illuminance	the amount of light received by a surface, measured in lux.
Impact transmission	sound produced and transmitted by an impact on the building fabric, for example, footsteps on a floor.
Infiltration	the movement of air through the fabric of a building.
Interstitial condensation	condensation which occurs within building elements.
Intrusive sound	sound reaching a room or space from an adjoining space or from outside the building. Typically it is a nuisance.
Ionising radiation	radioactivity which results in atoms losing or gaining electrons and becoming ionised, (ie. positively or negatively charged).
Kappa-value (κ-value)	a measure of the heat capacity of a building element.
Latent heat	the energy required to change a material from solid to liquid or liquid to gas. It does not result in temperature change.
Linear thermal transmittance ψ-value, (psi-value)	rate of heat transfer through a junction between two building elements, or at the perimeter of an opening.
Lumens (lm)	the measure of light emitted by a source. A light source which radiated 1 cd in all directions would have a luminous flux of 12.6 lm.
Luminous flux	the total amount of light given out by a source, measured in lumens.
Luminous intensity	the strength of a light source, defined as the power of visible light emitted for a solid angle of one steradian. Measured in candela.
Lux (lx)	the measure of the amount of light received by a surface. 1 lx is = 1 lumen/m^2.
Metabolic gains	gains in heat resulting from the presence of people in a building.
Modelling	the balance of direct and diffuse light in a space. A good balance enhances perception of the space and of human faces.
Moisture content	the amount of water vapour in air. Expressed in grams per kilogram of dry air (g/kg).
Numerical modelling	computer-based simulation of heat flows through building elements and components. More accurate than the combined method, but slower.
Operative temperature	a temperature value which incorporates the radiant and air temperatures and the effects of air movement.
Partial pressure	in a mixture of gases that part of the total pressure which is produced by one of the gases. For example, water vapour in air.
Percentage saturation	the ratio of moisture content of a body of air to its moisture content at saturation. Similar to, but not identical to relative humidity.
Phase change materials (PCM)	composite materials which include a material which melts at a set temperature, typically just above normal room temperature.

Photometric data sheets	data sheets which describe the performance of a luminaire.
Pressure (of a gas)	the force exerted by its atoms and molecules as they collide with each other and with solid surfaces. Expressed in pascals (Pa) or kilopascals (kPa).
Psychrometric chart	a chart which graphically expresses the relationships between air, vapour pressure and temperature.
Psychrometric	connected with the physical and thermodynamic properties of gas/air mixtures.
Radioactive decay	the change of a radioactive element to another element as a result of charged particles being emitted.
Radioactivity	the emission of sub-atomic charged particles from a material.
Relative humidity	the vapour pressure in air at a given temperature, expressed as a percentage of the saturated vapour pressure at that temperature.
Reverberation	the lengthening of sound duration as sound waves reflected from a surface reaching a listener after those coming directly from the source.
Saturation vapour pressure (SVP)	the maximum possible vapour pressure at a given temperature.
SAP	the Standard Assessment Procedure (generally known by its abbreviation) is the methodology used by the UK Government to assess and compare the energy and environmental performance of dwellings
SBEM	the Simplified Building Energy Model is a computer program developed for the UK Government to assess a building's energy consumption.
Sensible heat	the heat which results in a change of temperature in a material. 'Sensible' in the sense of being sensed or felt.
Solar gain	an increase in energy resulting from solar radiation falling on a surface.
Specific heat capacity	a measure of the amount of energy required to raise the temperature of a material. Measured in J/kgK.
Stack effect	the vertical air flow produced by temperature differences in an air column and wind action.
Standard temperature and pressure	a defined temperature and pressure used when examining the behaviour of gases. Usually 0°C and 100 kPa.
Steady state analysis	a method of assessing interstitial condensation risk in a building element using monthly average data. Also known as the Glaser method.
Steradian	a solid angle which is roughly 1/12.6 of the surface of a sphere.
Surface factor (F)	the ratio between the range of heat flow out of a surface to the range of heat flow into the surface.
Temperature	the fundamental unit of measure of heat.
Temperature factor	a measure of the cooling which takes place at the surface of a building element.
Thermal bridging	the regular interruption of one material in a construction by another with a different thermal conductivity or thermal resistance.
Thermal conductivity	a measure of the rate of heat transfer through a solid material. Expressed in W/mK.
Thermal diffusivity	a measure of how rapidly heat travels in a material.
Thermal effusively	a measure of the thermal inertia of a material.
Thermal mass	the capacity of the building fabric to absorb heat.
Thermal resistance	a measure of the resistance to heat transfer in a specified thickness of a material or cavity. Expressed in m^2 K/W.
Thermal transmittance	see U-value.
Time factor	the time delay between the maximum flows into and out of an element's internal surface.

Transient analysis	a method of condensation risk analysis which examines moisture movement on an hourly basis.
Troposphere	the lowest layer of the earth's atmosphere.
U-value	the rate of heat transfer through a building element, expressed in W/m^2K. Also known as the thermal transmittance.
Vapour permeability	the measure of how rapidly water vapour can diffuse through a material. Measured in gm/MNs.
Vapour permeance	the rate at which water vapour can diffuse through a given thickness of a material. Measured in g/MNs.
Vapour pressure	the pressure exerted by water vapour in a volume of air, or in a material. Usually measured in kilopascals (kPa).
Vapour resistance	the degree to which a material resists the passage of water vapour m easured in MNs/g.
Vapour resistivity	the measure of a material's resistance to the diffusion of water vapour. Measured in MNs/gm.
Wet bulb temperature	air temperature measured using a bulb which is covered in wet muslin and moved rapidly through the air. With a corresponding dry bulb temperature it allows vapour pressure to be determined.
Wind load	the forces acting on a building which are produced by differences in air pressure.

Bibliography

General

CIBSE (2015), *Guide A: Environmental design* (CIBSE: London).

01 Heat

BS EN ISO 6946:2007. Building components and building elements – Thermal resistance and thermal transmittance – Calculation method.
BS EN ISO 10456:2007. Building materials and products – Hygrothermal properties – Tabulated design values and procedures for determining declared and design thermal values.
BS EN ISO 13370:2007. Thermal performance of buildings – Heat transfer via the ground – Calculation methods.
BRE (2006), *BR 443 Conventions for U-value calculations* (BRE: Garston).
BRE (2007), *BR 497 Conventions for calculating linear thermal transmittance and temperature factors* (BRE: Garston).
BRE (2006), *IP1/06 Assessing the effects of thermal bridging at junctions and around openings* (BRE: Garston).
BRE (2013), The Government's Standard Assessment Procedure for Energy Rating of Dwellings (BRE: Garston).
The Concrete Centre (2012), *Thermal mass explained* (MPC – The Concrete Centre: Camberley).
NHBC Foundation (2012), *Understanding overheating – where to start: An introduction for house builders and designers* (NHBC Foundation: Milton Keynes).

02 Air

BS 5935:1991. Code of practice for ventilation principles and designing for natural ventilation.
BRE (2007), *BR 211 Radon: Guidance on protective measures for new buildings* (BRE: Garston).
CIBSE (2000), *Mixed mode ventilation* (CIBSE: London).
DCLG (2010), The Building Regulations, Approved Document F, Ventilation (DCLG: London).

03 Moisture

BS 5250:2011. Code of practice for control of condensation in buildings.
BS EN ISO 13788:2012. Hygrothermal performance of building components and building elements –Internal surface temperature to avoid critical surface humidity and interstitial condensation – Calculation methods
CIBSE (2012), *KS20 Practical psychrometry* (CIBSE: London).

04 Sound

DCLG (2004), The Building Regulations, Approved Document E, Resistance to the passage of sound (DCLG: London).
DfE (2015), *BB93 Acoustic design of schools: performance standards* (DfE: London).

05 Light

BS 8206-2:2008. Code of practice for daylighting.
Society of Light and Lighting (2012), *The SLL Code for Lighting* (SLL: London).
Society of Light and Lighting (2014), *LG10 Daylighting – a guide for designers,* (SLL: London).

Endnotes

01

1 Detailed discussion of this topic is beyond the scope of this book. Further information can be found in CIBSE (2013), *TM52: The Limits of Thermal Comfort: Avoiding Overheating in European Buildings* (CIBSE: London).
2 In Australia and New Zealand, where the range of temperatures is sufficient to affect the performance of thermal insulation, the resistance of the fibrous insulant material is reduced by a certain percentage for every degree above 23°C. The reduction depends on the insulation type.
3 Some countries express the thermal performance of an element in terms of the total thermal resistance, or R-value. This has the advantage of being simple to work with (adding a material with a certain resistance will increase the total resistance by that amount), but any R-value must be converted to a thermal transmittance in order to calculate heat transfer rates for building elements.
4 In the UK the application of the combined method for building regulations is governed by BRE (2006) *BR 443 Conventions for U-value calculations* (BRE: Garston).
5 While the 100-mm limit is suitable for assessing daily variations in temperature, a 20-mm depth limit is suitable for assessing hourly variations, and a 250-mm limit for weekly variations.
6 Occasionally people forget that thermal insulation does not just 'keep heat in'. Rather, it reduces the rate of heat transfer, whichever way the temperature gradient runs.
7 The UK's National Calculation Methodologies, iSBEM and SAP, work on a monthly basis. Strictly, they are quasi-steady state, because they take some account of the effect of thermal mass.

02

1 The Eurocodes, which contain the structural calculation methods, now refer to orography rather than topography. Orography deals with the formation and features of mountains and hills.
2 The stack effect in the first skyscrapers led to the introduction of the revolving door, because the differential pressures generated in the tall buildings made it difficult to operate conventional doors.
3 Other forces acting on the air can also damage buildings: air turbulence produced by aircraft can damage the roof coverings of buildings near the ends of runways.
4 Passivhaus is an energy efficiency standard promulgated by the Passivhaus Institut based in Germany. To qualify, a house must have a heating load of less than 10 W/m², provided by heating incoming fresh air. Meeting that standard requires low fabric U-values, high-performance windows and airtight construction.
5 One notable exception is the use of battens to form cavities between a vapour control layer and the internal surface on framed constructions: services can be run within the cavities, maintaining the integrity of the vapour control layer.
6 Rates of smoking have also dropped, from over 40% of females and 60% of males in 1950 to under 20% of the whole adult population in 2014. Smokers became a minority (albeit a sizeable one) in the latter half of the 1970s.
7 RCE-11 Radon and Public Health (Health Protection Agency, 2009).
8 UK requirements are set out in: Approved Document C (England, Wales); Section 3 of the Technical Handbooks to the Building Regulations (Scotland); Technical Booklet F (Northern Ireland).
9 For further guidance see CIBSE (2000), *AM13 Mixed mode ventilation* (CIBSE: London).

03

1 See CIBSE (2012), *KS20 Practical psychrometry* (CIBSE: London), page 12.
2 An explanation of the conversion factor can be found in BS 5250:2011, Appendix E.8.
3 Rising damp is surprisingly controversial. Over-diagnosis of rising damp by companies offering remedial treatment has produced 'rising damp deniers', who believe every problem labelled as rising damp will prove, on proper examination, to have another cause. We can be reasonably certain that most problems identified as rising damp have other causes, but there will be a minority of cases which are true rising damp. BR 466 Understanding dampness (BRE. 2004) estimated that in 5% of damp dwellings the primary cause was rising damp.
4 The method is described in BS EN ISO 13788.
5 See CIBSE (2009), *TM 48 The Use of Climate Change Scenarios for Building Simulation: the CIBSE Future Weather Years*, (CIBSE: London).

04

1. The same process of transmission can also result in vibrations from machinery and plant being transmitted through the fabric to where they can be sensed by the building occupants. Although not harmful, such vibrations can be a nuisance. For guidance see CIBSE (2002), *Guide B Heating, ventilating, air conditioning and refrigeration* (CIBSE: London), chapter 5.

05

1. For comprehensive (and rather daunting) guidance on the principles of lighting design see the Society of Light and Lighting (2012), *The SLL Code for Lighting* (SLL: London).
2. *The SLL Code for Lighting* contains over 50 tables for lighting conditions for different spaces, including autopsy rooms and mortuaries.
3. *The SLL Code for Lighting* plaintively observes: 'The reflectance of surfaces is often a parameter that is outside the control of the lighting designer; however, where possible, the designer should try to persuade those responsible to aim for reflectances in the above range.'
4. For a more detailed discussion, see Appendix B of BS 8300:2009+A1:2010: Design of buildings and their approaches to meet the needs of disabled people – Code of practice.
5. For further guidance on daylighting see BS 8206-2:2008. Code of practice for daylighting (which also addresses the amenity value of the view provided by windows) and the Society of Light and Lighting (2014), *LG10 Daylighting – a guide for designers* (SLL: London).

Index

Page numbers in *italics* denote figures and in **bold** denote tables.